INSIGHTS INTO CALCULUS

WITH THE

GRAPHICS CALCULATOR

HERBERT A. HOLLISTER

BOWLING GREEN STATE UNIVERSITY

D. C. Heath and Company

Lexington, Massachusetts / Toronto

Address editorial correspondence to:

D. C. Heath and Company
125 Spring Street
Lexington, MA 02173

**In memory of
Sandy**

Printed in the United States of America.

International Standard Book Number: 0-669-28905-1

10 9 8 7 6 5 4 3 2 1

PREFACE

Graphs play an important role in providing understanding of calculus and in applying its power. The graphics calculator magnifies this contribution by making graphs of functions quickly and simply available.

This text provides an introduction to the use of these instruments and illustrates some of the mathematical techniques that are now available because of technology. Furthermore, it provides greater insight into traditional topics through graphic illustration and simple mathematical programming. In no way should the discussions in this book be viewed as substitutes for the development of concepts in a calculus text; they are extensions of these ideas, not replacements.

The book is designed to be studied with paper, pencil, and graphics calculator at hand. Each example should be duplicated as it is read so that the student understands and can duplicate each step of the process.

The text is written with reference to the TI-81 calculator but almost all of the techniques can be applied using any one of several excellent instruments. You may need to translate from one form of notation to another but this should not present any serious barriers.

Thanks to the following reviewers for their valuable suggestions: Jim Jones, California State University-Chico; Roger Nelson, Lewis and Clark College; and Edward Norman, University of Central Florida.

Thanks to Ann Marie Jones and the staff of D. C. Heath and Company for their assistance and Carolyn Hollister for her help in the preparation of the manuscript.

H. A. Hollister

CONTENTS

1

INTRODUCTION

A certain degree of familiarity with your graphics calculator is necessary before attempting to use it in calculus.   The owner's manual provides the basic instruction for your particular instrument and you should be able to perform routine tasks with it and be familiar with its capacities. This section includes a brief review of some of the concepts and notation to be used as well as a few fundamental techniques that will strengthen your background and build additional skill.   As always, the notation used is that of the TI-81 but the techniques can be used with any programmable graphics calculator.

The RANGE settings determine the portion of the graph of the function that you will be able to view. The TI-81 displays the Standard setting as in Figure 1-1.

FIGURE 1-1

Minimum X-value is -10	Xmin=-10
Maximum X-value is 10	Xmax=10
X-scale is 1 unit per mark	Xscl=1
Minimum Y-value is -10	Ymin=-10
Maximum Y-value is 10	Ymax=10
Y-scale is 1 unit per mark	Yscl=1
X resolution: 1 pt per pixel	Xres=1

This means that a portion of the graph is drawn for $-10 \leq x \leq 10$ and $-10 \leq y \leq 10$ with both the x and y scales set at 1 and the x resolution is also 1. In this book, we will list a RANGE setting by giving the appropriate sequence of numbers rather than writing the abbreviations each time. Thus, for example, the Standard RANGE setting is [-10,10,1,-10,10,1,1]. The Trig RANGE setting is [-2π,2π,π/2,-3,3,.25,1]. (Decimal approximations are used for π.)

Example 1. Draw the graph of

$$y = x^2 - 3x + 5$$

SOLUTION: Press $\boxed{Y=}$ and the screen is displayed as in Figure 1-2.

FIGURE 1-2

Enter the function as

$$Y_1 = X^2 - 3X + 5$$

Set the RANGE setting at "Standard" ($-10 \leq x \leq 10$ and $-10 \leq y \leq 10$) and press GRAPH.   The screen will look like Figure 1-3.

FIGURE 1-3

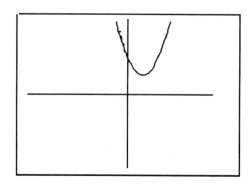

A closer view of the graph can be obtained by using ZOOM IN or changing the RANGE setting. Using TRACE to move the cursor to the vertex and then applying ZOOM IN yields Figure 1-4.

FIGURE 1-4

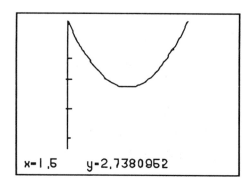

2

For an even closer view of part of the graph, we can use TRACE to move the cursor to the portion of the graph we would like to examine and then ZOOM IN again as in Figure 1-5.

FIGURE 1-5

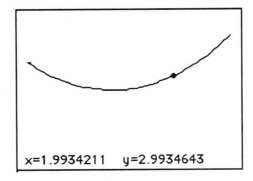

x=1.9934211   y=2.9934643

Notice that the x and y values are displayed on the screen as we move the cursor. ‖

We will leave the ZOOM factor at 4 for all our examples.

You can draw several graphs at once on the TI-81; just list them under Y= as Y_1, Y_2, Y_3, and Y_4. If you only want the graph of one function, you may list it as any of the four. Pressing the GRAPH key will give you the display. You may choose to draw all graphs at once or in sequence. Most of the time you will get more insight by drawing them sequentially.

Example 2. Draw the graphs of

$$y = x^2, \qquad y = x^3, \qquad y = \frac{4}{x}, \quad \text{and} \quad y = \frac{6}{x^2+1}$$

on one coordinate system with the Standard RANGE setting.

SOLUTION: List the functions as
$$Y_1 = X^2$$
$$Y_2 = X^3$$
$$Y_3 = 4/X$$
$$Y_4 = 6/(X^2+1)$$

The graph appears in Figure 1-6.

FIGURE 1-6

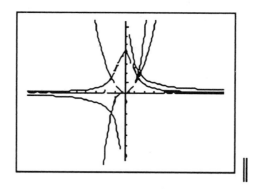

SHIFTS

When we shift a curve from one location to another on a coordinate system, the new curve is congruent to the original, that is it has all the same geometric properties.   The idea of shifting graphs is illustrated quite well using the graphics calculator.

Example 3. Draw the graphs of

$$y = x^2 \qquad and \qquad y = x^2 + 2$$

on the same coordinate system.   Do they appear to be congruent?

SOLUTION: The graphs are given in Figure 1-7.

FIGURE 1-7

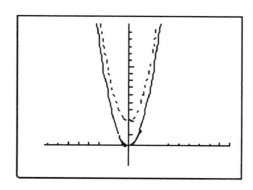

The two curves have the same shape.   The second is just the first shifted upward 2 units and, therefore, they are congruent. ‖

Example 4. Draw the graphs of

$$y = x^2 \qquad \text{and} \qquad y = (x - 3)^2$$

on the same coordinate system. Do they appear to be congruent?

SOLUTION: The graphs are given in Figure 1-8.

FIGURE 1-8

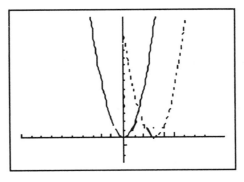

The curves are congruent; the second is the first shifted 3 units to the right. ‖

Example 5. Draw the graphs of

$$y = x^2 \qquad \text{and} \qquad y = (x - 3)^2 + 2$$

on one coordinate system. Do the graphs appear to be congruent?

SOLUTION: The graphs are given in Figure 1-9.

FIGURE 1-9

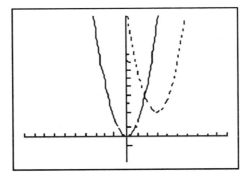

The two curves are congruent; the second is the first shifted 3 units to the right and 2 units upward. ‖

IN GENERAL:

The graphs of
$$y = f(x) \quad \text{and} \quad y = f(x - b) + c$$
are congruent.

The graph of
$$y = f(x - b) + c$$
is the graph of
$$y = f(x)$$
shifted $|b|$ units along the x-axis (to the right if b is positive and to the left if b is negative) and $|c|$ units along the y-axis (up if c is positive and down if c is negative).

Example 6. The graphs of
$$y = x^3 \qquad \text{and} \qquad y = (x - 2)^3 + 1$$
are congruent; the second is the first shifted 2 units to the right and 1 unit upward.   The graphs are given in Figure 1-10,.

FIGURE 1-10

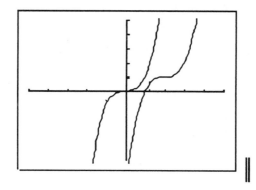

Shifts of curves are sometimes treated under the topic of Translations of Axes but we will not explore such a formal development here.

SOLVING EQUATIONS

A   graphing calculator can be used to find approximate solutions to equations that do not yield to the usual techniques of algebra.   We draw the graph of a function that is naturally associated with the equation and estimate the solutions by estimating the points where the graph crosses the x-axis. There are two basic techniques.

GRAPHICS METHOD

Example 7.   Solve the equation
$$x^3 + 2x^2 - 5x - 2 = 0$$

SOLUTION: We let
$$y = x^3 + 2x^2 - 5x - 2$$

and draw the graph in Figure 1-11 using the Standard RANGE setting.

FIGURE 1-11

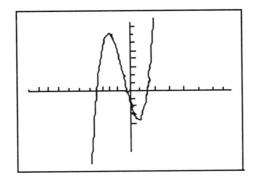

From the graph we see that there are at least three zeros and basic
algebra assures us that a polynomial of degree three can have no more
than three zeros.

If we use the TRACE key and move the cursor to the right, the graph in
Figure 1-12 indicates that the curve crosses the x-axis for x near 1.79.

FIGURE 1-12

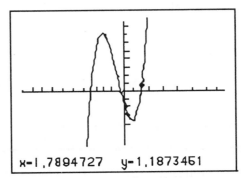

x=1.7894727   y=1.1873451

By applying ZOOM IN and using TRACE in the first drawing in
Figure 1-13 we see that a zero occurs near 1.65. Another application
of ZOOM IN produces the second graph in Figure 1-13 and using
TRACE again yields y = -0.0376561 when x =1.6776316.

If this is close enough for our purpose, we quit at this point.

FIGURE 1-13

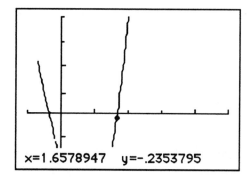
x=1.6578947   y=-.2353795

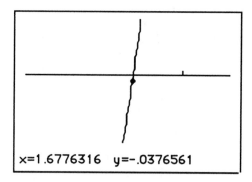
x=1.6776316 y=-.0376561

If this approximation is not sufficient, we can use ZOOM IN and TRACE again as in Figure 1-14 and see that y = 0.012661657 when x = 1.6825658.

FIGURE 1-14

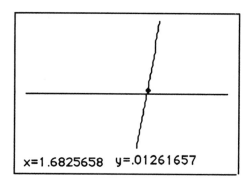
x=1.6825658 y=.01261657

At this point we would say that x = 1.6825658 is a sufficient approximation for a solution.

Approximations of the other zeros can be found in the same manner. ▌

COMPUTATIONAL METHOD

If we want or need a better approximation than we obtained in Example 7, we can combine information read from the graph with a program we call **ZERO**. The program is based on the following fact.

If the graph of f(x) has no breaks on the interval [a,b], is positive at one endpoint, and is negative at the other endpoint, then f(c) = 0 for some number c between a and b.

This is a special case of the Intermediate Value Theorem that you will encounter in calculus and leads to the following technique.

The Bisection Method for approximating zeros of a function:

If the graph of f(x) has no breaks on [a,b] and f(a)f(b) < 0:

1. Let m = (a+b)/2, the midpoint of [a,b]. This is the first approximation of a zero.

2. If f(a) and f(m) agree in sign, then the zero is between m and b. If f(m) and f(b) agree in sign, then the zero is between a and m.

3. The second approximation is the midpoint of whichever of these two new intervals contains the zero.

4. Now repeat the process as many times as is necessary until the length of the new interval is less than your required degree of accuracy. The midpoint of this interval is a zero of the function within the degree of accuracy specified.

The method is illustrated in Figure 1-15.

FIGURE 1-15

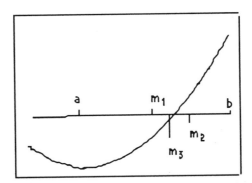

The bisection method can be stored in your calculator with the following program called ZERO. The notation is that of the TI-81 but it can be adapted to any programmable calculator.

The **ZERO** program:

```
:Pgm :ZERO
:Disp"CHOOSE A,B SO F(A)F(B)<0"
:Lbl 1
:Disp"ENTER A"
:Input A
:Disp"ENTER B"
:Input B
:.000000001→E
:A→X
:Y₁→P
:B→X
:If P*Y₁>0
:Goto 8
:Lbl 2
:(A+B)/2→M
:A→X
:Y₁→P
:M→X
```

```
:If abs(A-B)<E
:Goto 9
:If P*Y₁>0
:Goto 7
:M→B
:Goto 2
:Lbl 7
:M→A
:Goto 2
:Lbl 9
:Disp"ZERO AT"
:Disp X
:Disp "Y₁="
:Disp Y₁
:Goto 5
:Lbl 8
:Disp"F(A)F(B)>0, CHOOSE AGAIN"
:Goto 1
:Lbl 5
:End
```

In order to use the program you first need to enter the function $F(X)$ as Y_1 and draw the graph. Use the graph to find two numbers A and B such that $F(A)$ and $F(B)$ differ in sign and there is just one zero of the function between A and B.   Then apply the program.

Example 7. (revisited)

SOLUTION: Looking again at Figure 1-11, we see that there is one zero somewhere between -4 and -3, one between -1 and 0, and one between 1 and 2.

The results of three applicatons of the **ZERO** program are displayed in the screens of Figure 1-16.

FIGURE 1-16

```
?-4              ?-1              ?1
ENTER B          ENTER B          ENTER B
?-3              ?0               ?2
ZERO AT          ZERO AT          ZERO AT
    -3.323404276     -.3679263673     1.681330644
Y₁=              Y₁=              Y₁=
       -2.88E-9         -1.221E-9         4.67E-9
```

The program yields a zero at x = 1.681330644 with
y = 0.00000000942. Using the program took no more time than
ZOOMING IN and gave better accuracy.

The other zeros are at x = -3.323404276 and x = -0.357963673.

The values of Y_1 are displayed just to be sure that the program was used
correctly. ‖

The **ZERO** program has a built in allowable error of E = .000000001.
If you want a different degree of accuracy, just edit the program and replace E
by your specification.

This program can also be used to solve some systems of two equations by
determining the points where two graphs intersect.

Example 8. Solve the system of equations
$$y = x^4 - 9x + 5$$
$$y = x^3 + x + 7$$

SOLUTION: First we draw Figure 1-17 letting
$$Y_2 = X^4 - 9X + 5$$
$$Y_3 = X^3 + X + 7$$
and $Y_1 = (X^4 - 9X + 5)-(X^3 + X + 7)$

The zeros of Y_1 will be the x-coordinates of the points of intersection.

We will use the calculator to approximate these and the corresponding
y-coordinates for Y_2 and Y_3.

FIGURE 1-17

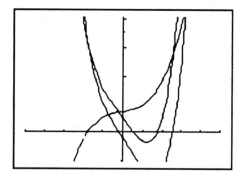

By drawing the graphs in sequence, we see that there appear to be zeros of Y_1 between -1 and 0 and between 2 and 3. The results are displayed by the four screens in Figure 1-18.

FIGURE 1-18

```
?-1
ENTER B
?0
ZERO AT
        -.1990542994
Y1=     -2.876E-9
```

```
?2
ENTER B
?3
ZERO AT
        2.596954236
Y1=     -8.92E-9
```

```
Y2
        6.79305864B

Y3
        6.79305864D
```

```
Y2
        27.11125847

Y3
        27.1112584B
```

The solutions of the system are approximated by (-0.1990542994,6.79305864) and (2.596954236,27.1112584). ‖

In Example 8, we computed the y-coordinates by using the value of the zero that was stored in X to let the calculator compute the corresponding values of Y_2 and Y_3. By setting up your problem carefully, you can frequently simplify the computational work that follows the graphing.

3.   Draw the graphs of the following functions on one coordinate system.

$$y = x^4, \quad y = x^4 - 16, \quad y = x^4 - x^2, \quad y = x^4 - 3x^2 - 4$$

(Start with a RANGE setting of $[-5,5,1,-20,20,4,1]$.)

What are the x-intercepts for each?

4.   Draw the graphs of the following functions on one coordinate system.

$$y = 1/(x^2-1), \quad y = x/(x^2-1), \quad y = x^2/(x^2-1), \quad y = x^3/(x^2-1)$$

Try this problem with the range setting from Exercise 3 and then try it with $[-4,4,1,-4,4,1,1]$.   Which gives the better illustration? What are the x-intercepts?   What do the graphs have in common?

5.   Draw the graphs of the following functions on one coordinate system.

$$y = (x^2+1)/x, \quad y = (x^2+1)/x^2, \quad y = (x^2+1)/x^3, \quad y = (x^3+1)/x^3$$

(Use the $[-4,4,1,-4,4,1,1]$ setting.)   Are there any x-intercepts?   What do the graphs have in common?

6.   Draw the graphs of the following functions on one screen.

$$y = x^2 - 3x, \quad y = x^2 -3x +1, \quad y = x^2 - 3x - 2, \quad y = x^2 -3x + 4$$

How many x-intercepts for each?   What do the graphs have in common? Justify your conclusion.

7.   Draw the graphs of

$$y = \frac{5}{x} \quad \text{and} \quad y = \frac{5}{x - 1}$$

on the same coordinate system.   Are the curves congruent?   Why?

8.   Draw the graphs of

$$y = \frac{x^3}{(x - 2)} \quad \text{and} \quad y = \frac{(x + 2)^3}{x}$$

on the same coordinate system.   Are the graphs congruent?   Why?

9.   Draw the graphs of

$$y = 2x^2 \quad \text{and} \quad y = 2(x - 1)^2 + 3$$

on the same coordinate system.   Are the curves congruent?   Why?

10. Draw the graphs of

$$y = \frac{x + 1}{x + 3} + 2 \quad \text{and} \quad y = \frac{x - 2}{x + 1} - 1$$

on the same coordinate system. Are the graphs congruent? Why?

11. (a) Draw the graphs of

$$y = x^2 \qquad \text{and} \qquad y = x^2 + 4x + 5$$

on one coordinate system. Do the curves appear to be congruent? Use the facts given in this section to justify your conclusion. (Hint: Complete the square.)

(b) Choose any numbers b and c and draw the graphs of

$$y = x^2 \qquad \text{and} \quad y = x^2 + bx + c$$

on one coordinate system. Do the curves appear to be congruent? Does the shape of the curve depend on b and c? Justify your conclusion.

12. Draw the graphs of

$$y = 2x^2 \qquad \text{and} \qquad y = 2x^2 - 12x + 18$$

on one coordinate system. Are the curves congruent? Use the facts given in this section to justify your conclusion.

13. (a) Draw the graphs of the following functions on one coordinate system.

$$y = x^2 \qquad y = 2x^2 \qquad y = -3x^2 \qquad y = (2/3)x^2$$

What do the four graphs have in common?

(b) Choose a function $f(x)$ other than x^2 and a number $a \neq 1$, $a \neq 0$. Draw the graphs of $y = f(x)$ and $y = af(x)$ on one coordinate system.

(c) If $f(x)$ is any function and a is any number, what do your results from parts (a) and (b) lead you to say about the graphs of

$$y = f(x) \qquad \text{and} \quad y = af(x)?$$

14. (a) Choose a function $f(x)$ and numbers a, b, c with $a \neq 1$, $a \neq 0$ and draw the graphs of
$$y = f(x), \quad y = af(x), \quad y = af(x - b), \quad \text{and} \quad y = af(x - b) + c$$
on one coordinate system.

   (b) If $f(x)$ is any function and a, b, c are any numbers, what do your results from part (a) lead you to say about the graphs of
$$y = f(x), \quad y = af(x), \quad y = af(x - b), \quad \text{and} \quad y = af(x - b) + c$$

15. Use the Graphics Method to approximate a solution of
$$x^3 + 9x - 3 = 0$$
to within two decimal places.

16. Use the Graphics Method to approximate the solutions of
$$x^5 - 7x^4 - 2x^3 + 3x^2 + 7x - 4 = 0$$
to within two decimal places.

17. Use **ZERO** to approximate solutions of
$$x^4 - 10x^3 - 40x^2 + 19x + 13 = 0$$

18. Use **ZERO** to approximate solutions of
$$x^3 + 5x^2 - 3x + 4 = 2x^2 - 3x + 9$$

19. Use **ZERO** to approximate solutions of
$$\frac{x^3}{x^2 + 7} = \frac{2x + 3}{x^2 + 1}$$

20. Use **ZERO** to approximate solutions of
$$(x - 5)^4 = (2x + 5)^3 - (3x - 1)^2$$

21. Use **ZERO** to approximate solutions of the system
$$y = x^3 + 5x - 7$$
$$y = x^2 - 5$$

22. Use **ZERO** to approximate solutions of the system

$$y = \frac{x^3 + x - 11}{x^2 + 8}$$

$$y = \frac{2x + 2}{x^2 + 4}$$

23. Use graphs to show that

$$y = f(x) = .5x^2 - 3x + 2$$

is neither even nor odd.

24. Use graphs to decide whether

$$y = f(x) = 3x^4 - x^2 + 5$$

is even, odd, or neither.

25. Use graphs to decide whether

$$y = f(x) = 2x^3 + 3x$$

is even, odd, or neither.

26. Use graphs to decide whether

$$y = g(x) = x^2 - x^{-2}$$

is even, odd, or neither.

27. Use graphs to decide whether

$$y = g(x) = x^{-3} - 4x$$

is even, odd, or neither.

28. Use graphs to decide whether

$$y = f(x) = \frac{6}{x^2 + 1}$$

is even, odd, or neither.

29. Use graphs to decide whether

$$y = f(x) = \frac{x}{x^3 - 1}$$

is even, odd, or neither.

30. Use graphs to decide whether

$$y = g(x) = \frac{x^2 - 7}{x^3}$$

is even, odd, or neither.

31.   Use graphs to decide whether
$$y = g(x) = \frac{x^4 + x^2 + 2}{3x^2 - 1}$$
is even, odd, or neither.

EXTENSION

If $f(x)$ is any function such that $f(-x)$ has meaning whenever $f(x)$ has meaning, we can use $f(x)$ to define two special functions
$$E_f(x) = \frac{f(x) + f(-x)}{2} \quad \text{and} \quad O_f(x) = \frac{f(x) - f(-x)}{2}$$

A simple algebraic verification shows that $E_f(x)$ is even, $O_f(x)$ is odd, and that

$$f(x) = E_f(x) + O_f(x)$$

The function $E_f(x)$ is called the *even part* of $f(x)$ and $O_f(x)$ is called the *odd part*.   Thus we see that if $f(-x)$ has meaning whenever $f(x)$ does, $f(x)$ can be expressed as the sum of an even function and an odd function.   In fact, this is the *only* way that $f(x)$ can be expressed as the sum of an even and odd function.

32.   Determine the even and odd parts of
$$y = .4x^3 + 2x^2 - 3x + 1$$
and draw the graphs of all three functions.

33.   Determine the even and odd parts of
$$y = \frac{x^3}{x^2 + 9}$$
and draw the graphs of all three functions.

34.   Determine the even and odd parts of
$$y = \frac{x^2 + x - 4}{x^2 + 1}$$
and draw the graphs of all three functions.

2

ALGEBRAIC FUNCTIONS

An algebraic function is one that can be expressed as a finite number of sums, differences, multiples, quotients, and radicals involving x^n. The following are examples of algebraic functions.

$$f(x) = x^3 - 7x + 3 \qquad g(x) = \sqrt{x^4 + 5x^2 + 19}$$

$$p(x) = \frac{3x^5 + 2x + 43}{5x^2 - 9} \qquad q(x) = \sqrt[3]{\frac{x^{.5} + 9}{2x^{12} - 4x^2 + 3}}$$

A graphics calculator can be used to draw graphs of algebraic functions and give special insight into their properties.

POLYNOMIAL FUNCTIONS

Among the first functions you studied were those that can be expressed as polynomials. Not surprisingly, these are called polynomial functions. Even though they are relatively simple, they do have some very special and important properties.

Example 1. The functions

$$y = x^2 - 3x + 7 \qquad y = x^3 - 5x^2 + 2x - 4$$

are polynomial functions. Their graphs are given in Figure 2-1.

FIGURE 2-1

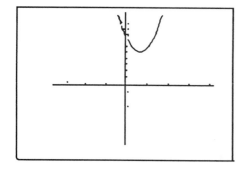 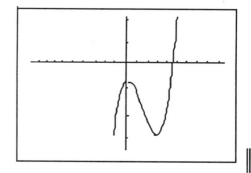

Before continuing in your reading, please choose any three polynomial functions of odd degree and draw their graphs on separate coordinate systems.

Did each of your graphs cross the x-axis? If not, use TRACE and shift each graph to the left or right until it does so. Your graphs and those in Figure 2-1 illustrate an important property of polynomial functions.

1. A polynomial function of odd degree has at least one zero.

The next property is usually established in basic algebra.

2. A polynomial function of degree n has no more than n zeros.

Graphically, this means that the graph of a polynomial function of degree n cannot cross the x-axis more than n times.

Now examine the graphs in Figure 2-1 to see how many times each graph changes direction: up from down or down from up. See if you can create a polynomial function of degree 4 that changes direction three times.

These observations illustrate a third property:

3. The graph of a polynomial function of degree n can change direction no more than n-1 times.

Example 2. The graph of
$$y = x^4 - 4x^2 + x + 2$$
is drawn in Figure 2-2 with the Standard RANGE setting.

FIGURE 2-2

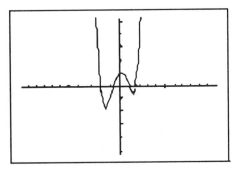

It is of degree 4 and changes direction exactly 3 times. ‖

Example 3. The graph of
$$y = x^4 + 3x^2 - 2$$
is given in Figure 2-3.

FIGURE 2-3

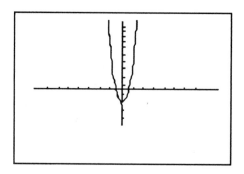

It has only one change in direction. ‖

For any polynomial function, as x increases in value, eventually the term with the highest power of x dominates the function. This term is called the *leading term*.

Example 4. The function
$$y = 2x^3 - 7x^2 + 3x + 5$$
is dominated by the leading term
$$2x^3$$
because as x gets large, $2x^3$ is much larger in absolute value than any other term.

For instance, for $x = 100$,
$$2x^3 = 2000000$$
while
$$-7x^2 = -70000$$
and
$$3x = 300$$

The graphs of
$$y = 2x^3$$
and
$$y = 2x^3 - 7x^2 - 3x + 5$$
are given in Figure 2-4. The RANGE setting is [100,101,.1,1900000,2000000,10000,1]. The two graphs appear to be the same for these large values of x.

FIGURE 2-4

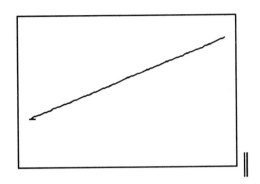

The graph of a polynomial function with even degree and a positive coefficient in the leading term will become arbitrarily large for values of x that are large in absolute value. If the coefficient is negative, then the function values will be negative but large in absolute value.

Similarly, the values of a polynomial function of odd degree will get large in absolute value as x gets large in absolute value but they will be opposite in sign. Since there are no breaks in the curve, a polynomial of odd degree will always have at least one zero. The methods of Section 1 can be used to approximate these zeros.

Example 5. Approximate any zeros of
$$f(x) = x^3 - 4x^2 + 8x - 6$$

SOLUTION: The graph is given in Figure 2-5.

FIGURE 2-5

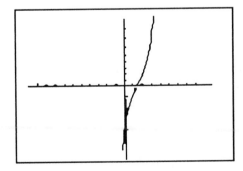

From the graph there appears to be only one zero and it is between 1 and 2. We use the **ZERO** program to approximate it; the results are displayed in Figure 2-6.

FIGURE 2-6

```
?1
ENTER B
?2
ZERO AT
        1.361103081
Y₁=
            1.64E-9
```

The only zero is x = 1.361103081. ▌

ASYMPTOTES

The graph of a function may approach a line.   When this happens, the
line is called an *asymptote* for the curve or the function.

1. The line x = a is a vertical asymptote for the function f(x)
    if the function values get large (or get large in absolute value
    but negative) for values of x on one side of a but close to a.
2. The line y = b is a horizontal asymptote for the function f(x)
    if the function values get close to b as x gets large or the
    absolute value of x gets large but x is negative.
3. The line y = mx+b is an asymptote for the function f(x) if
    [f(x) - (mx+b)] gets close to 0 as x gets large or as the
    absolute value of x gets large but x is negative.

There are times when a portion of a graph actually seems to be the line
because the graphs are closer together than the width of the curve.

Example 6. Draw the graph of

$$y = \frac{2}{x^2 - 1}$$

and determine any asymptotes.

SOLUTION: The graph is given in Figure 2-7.   The RANGE setting is
[-4,4,1,-4,4,1,1].

FIGURE 2-7

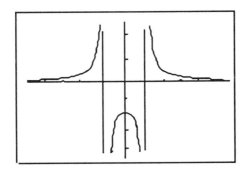

The curve gets close to the vertical line $x = 1$ as x gets close to 1. It also gets close to the vertical line $x = -1$ as x gets close to -1. For values of x with large absolute values, the graph gets close to the line $y = 0$, the x-axis. ‖

Example 7. Draw the graph of

$$y = \frac{x^2}{x^2 + 1}$$

and determine any asymptotes.

SOLUTION: The graph is given in Figure 2-8 with a RANGE of $[-20, 20, 5, -1, 3, 1, 1]$.

FIGURE 2-8

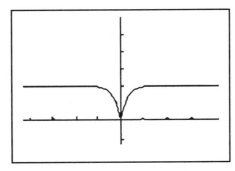

The graph gets close to the line $y = 1$ as $|x|$ gets large. ‖

Example 8. Draw the graph of

$$y = \sqrt{x^2 - 1}$$

and determine any asymptotes.

SOLUTION: The graph is given in Figure 2-9.

FIGURE 2-9

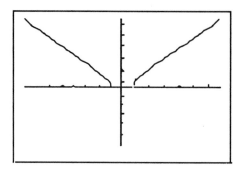

The curve gets close to the line $y = x$ for large values of x and close to the line $y = -x$ for negative values of x with large absolute value. ‖

Your graphics calculator may display vertical asymptotes if the RANGE setting provides sufficient space. Figure 2-10 shows two graphs of

$$y = \frac{9x+3}{x^2-1} \, ,$$

the first with $-20 \leq x \leq 20$ and the second with $-2 \leq x \leq 2$. The asymptotes appear in the second drawing.

FIGURE 2-10

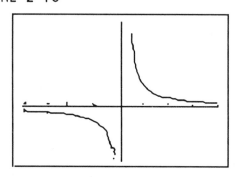 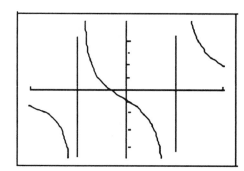

If you draw a graph that seems to have a vertical asymptote at a point, you can use TRACE and ZOOM IN to get a better view of the graph near the value of x in question.

Example 9. The graph of

$$y = \frac{1}{x^2 - 2x - 5}$$

is given twice in Figure 2-11. We use the Standard RANGE setting for the first drawing, The asymptote appears when we use TRACE to move the x-coordinate to about -1.5, and then ZOOM IN.

FIGURE 2-11

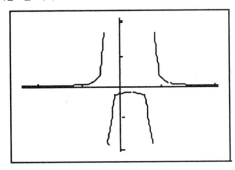 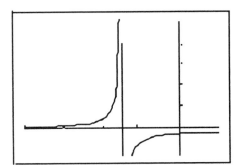

Horizontal asymptotes may not be as obvious with your calculator because you need to observe the function values as x moves away from zero. Sometimes you can get a better idea of the behavior by using TRACE and moving the x values to the right or to the left.

Example 10. Determine any horizontal asymptotes of

$$y = \frac{9x^2}{x^2 + 5}$$

SOLUTION: The first drawing in Figure 2-12 does not indicate any asymptotes. In the second drawing we use TRACE to move the x-coordinate to values larger than 20 and the y-coordinates seem to be getting close to 9. The line y = 9 appears to be a horizontal asymptote.

FIGURE 2-12

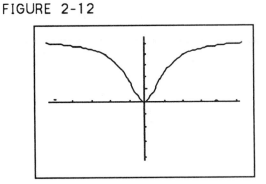 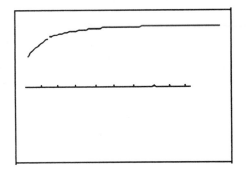

Further verification is provided in Figure 2-13 by using a RANGE setting of [-50,50,10,8,10,1,1] and setting Y_2 = 9. For large values of $|x|$, the curve and the line seem to coincide.

FIGURE 2-13

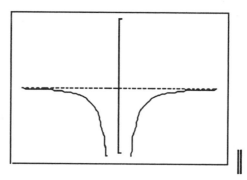

Finding asymptotes that are neither vertical nor horizontal is not quite as simple. In most cases it requires looking at the graph for a large interval of x-values.

Example 11. The graph of

$$y = \sqrt{4x^2 + 29}$$

is given in Figure 2-14, first drawn with the Standard RANGE setting and then with a setting of [0,80,10,10,300,50,1].

FIGURE 2-14

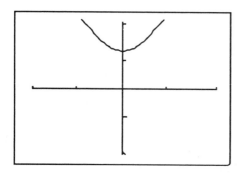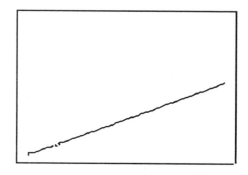

The second graph looks very much like a line. This is not really too surprising; we might expect $\sqrt{4x^2 + 29}$ to be close to $2x$ for large values of x. The line $y = 2x$ is, in fact, an asymptote for the curve. ‖

Notice that polynomial functions of degree two or more do not have asymptotes.

RATIONAL FUNCTIONS

A function is said to be *rational* if it can be expressed as the quotient of two polynomials.   The functions

$$y = \frac{7x}{x^2 + 5} \qquad f(x) = \frac{x^4 - 7x^2 + 5x - 7}{x^2 + x - 1} \qquad y = x^3 - 6x + 3$$

are examples of rational functions.   Polynomial functions are rational functions with denominator 1.

Rational functions that are not polynomial functions may not have meaning for some values of x, that is, the domain of the function may not be all the real numbers.   In fact, Examples 6 and 9 above are such functions.   In each of these cases, the function had vertical asymptotes at the points where the denominator would have been zero, the numerator non-zero, and the function undefined.   This property holds in general.

1. If the numerator and denominator of a rational function have no common factors, then the graph has a vertical asymptote at any zero of the denominator.

Example 12. The rational function
$$y = \frac{2x + 3}{x^2 - 3x - 4}$$

has vertical asymptotes at x = -1 and x = 4.   The graph is given in Figure 2-15.

FIGURE 2-15

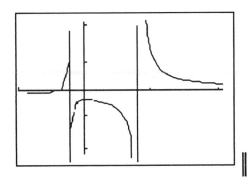

In Example 10, you saw that the graph of
$$y = \frac{9x^2}{x^2 + 5}$$
has a horizontal asymptote of $y = 9$. What about other rational functions?

To get better insight into the question, please use your calculator to draw graphs of the following functions and see if any of them appear to have horizontal asymptotes.

$$y = \frac{x^3}{x^3 + 7} \qquad y = \frac{4x^2 + 2x + 5}{x^2 - 2} \qquad y = \frac{7x^2 + 5x + 4}{x^3 - 3} \qquad y = \frac{x^4 + 2x + 5}{x^2 - 2x - 3}$$

From the graphs, it appears that the first three functions have horizontal asymptotes and the fourth does not.   The last function gets large as x gets large.

2. If the degree of the numerator and the denominator of a rational function are the same, then the line $y = b$ is a horizontal asymptote where b is the quotient of the leading terms.

3. If the degree of the numerator is less than the degree of the denominator, then the line $y = 0$ is a horizontal asymptote.

4. If the degree of the numerator is greater than the degree of the denominator, then there is no horizontal asymptote.

Example 13. Determine any asymptotes of
$$y = \frac{4x^2 - 3x + 2}{x^2 + 5x - 6}$$
and draw the graph.

SOLUTION: The lines
$$x = -6 \quad \text{and} \quad x = 1$$
are vertical asymptotes because the denominator is 0 at $x = -6$ and $x = 1$.

The line
$$y = 4$$

is a horizontal asymptote because the numerator and denominator are of the same degree and the quotient of the leading terms is **4**.

The graph is given in Figure 2-16

FIGURE 2-16

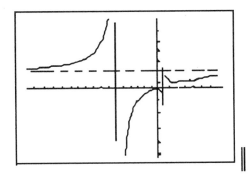

Example 14. Determine any asymptotes of
$$y = \frac{5x^3 - 3x^2 + 2x - 3}{x^4 - 16}$$

and draw the graph.

SOLUTION: The line
$$y = 0$$

is a horizontal asymptote because the numerator is of lower degree than the denominator.   The vertical asymptotes are
$$x = -2 \quad \text{and} \quad x = 2.$$

The graph is given in Figure 2-17.

FIGURE 2-17

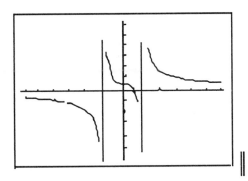

Example 15. Determine any asymptotes of
$$y = \frac{x^3 - 3x + 5}{7x^2 + 2}$$

and draw the graph.

SOLUTION: The function does not have a horizontal asymptote because the degree of the numerator is greater than the degree of the denominator. There are no vertical asymptotes because the denominator has no zeros. The graph is given in Figure 2-18.

FIGURE 2-18

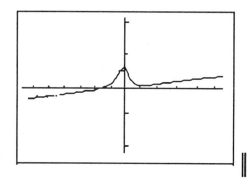

If you draw the graph of a rational function and it appears to have a vertical asymptote, you can approximate the x-value by approximating the zero of the denominator.

Example 16. Determine any asymptotes of
$$y = \frac{x^3 + 5x - 9}{x^3 + 7x^2 - 10}$$

SOLUTION: The graph is given in Figure 2-19.

FIGURE 2-19

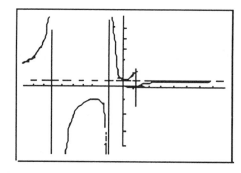

Since the degree of the numerator is the same as the degree of the denominator, the horizontal asymptote is the line $y = 1$. There seem to be three vertical asymptotes, one between -7 and -6, one between -2 and -1, and one between 0 and 4. We set Y_1 equal to the denominator

$$Y_1 = x^3 + 7x^2 - 10$$

and use ZERO to approximate these values of x. The results are displayed in Figure 2-20.

FIGURE 2-20

```
?-7
ENTER B
?-6
ZERO AT
        -6.782627785
Y1=
            -3.7E-9
```

```
?-2
ENTER B
?-1
ZERO AT
        -1.327770803
Y1=
            -4.18E-9
```

```
?0
ENTER B
?4
ZERO AT
        1.110398565
Y1=
            4.18E-9
```

The vertical asymptotes are the lines $x = -6.782627785$, $x = -1.327770803$, and $x = 1.110398565$. ‖

For values of x that are not near the vertical asymptotes, rational functions behave much like polynomial functions and most of this behavior is determined by the numerator. The denominator only affects the magnitude.

Example 17. The graphs of the functions

$$y = \frac{x^2 + x - 20}{x + 1} \quad \text{and} \quad y = \frac{x^2 + x - 20}{5} \quad \text{for} \quad 3 \le x \le 6$$

are given in Figure 2-21. We chose 5 for the second denominator because the denominator is near 5 for $3 \le x \le 6$.

FIGURE 2-21

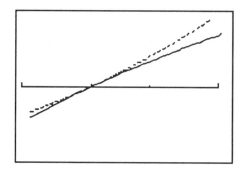

Notice that the two functions cross the x-axis at the same point. This is so because a fraction can be zero only if the numerator is zero. ▌

Algebraic functions other than rational functions can be very complicated. Since even roots of negative numbers have no meaning in the real number system, a particular algebraic function may not have meaning for large sets of numbers.

Example 18. The function
$$y = \sqrt{x^2 - 4x - 5}$$
will have meaning only when
$$x^2 - 4x - 5 \geq 0$$
The graph is given in Figure 2-22.

FIGURE 2-22

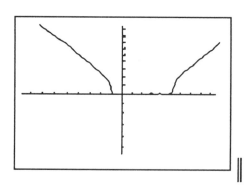

A function that involves even roots can have both vertical and horizontal asymptotes even though the function is not defined on some intervals.

Example 19. Draw the graph of

$$y = \frac{4}{\sqrt{x^2 - 9}}$$

and determine any asymptotes.

SOLUTION: The graph is given in Figure 2-23 with the Standard RANGE setting.

FIGURE 2-23.

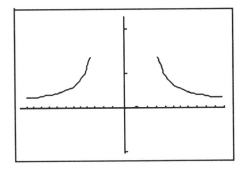

It appears that the x-axis and the lines x = -3 and x = 3 are the asymptotes. ‖

Notice in Example 19 that the function need not be defined on both sides of the x-value where a vertical asymptote occurs; all that is required is that the y-values get large as x approaches the value from one side.

EXERCISES

1. How many zeros might the function
    $$y = 5x^4 - 2x^2 - 7$$
    have? Use a graph to determine the actual number.

2. How many zeros might the graph of
    $$y = x^2 + 2x + 5$$
    have? Use a graph to determine the actual number.

3. How many zeros might the function
    $$y = x^6 - 5x^5 + 4x^3 + 2x^2 - 10x + 1$$
    have? Use a graph to determine the actual number.

4.   How many zeros might the function
$$y = 2x^5 + 4x^4 - 3x^3 + 2x + 7$$
have?   Use a graph to determine the actual number.

5.   How many times might the graph of
$$y = 3x^3 - 7x^2 + 5x - 3$$
change direction?   Use a graph to determine the actual number of changes.

6.   How many times might the graph of
$$y = x^5 - 10x - 3$$
change direction?   Use a graph to determine the actual number of changes.

7.   How many times might the graph of
$$y = x^3 + 4x - 7$$
change direction?   Use a graph to determine the actual number of changes.

8.   How many times might the graph of
$$y = x^4 - 4x^3 - 7x + 3$$
change direction?   Use a graph to determine the actual number of changes.

9.   Illustrate how the degree term of a polynomial function dominates the function by drawing the graphs of
$$y = x^3 - x^2 + x \qquad \text{and} \qquad y = x^3$$
on the same coordinate system with a RANGE setting of [-10,10,1,-1200,1200,200,1].

10.   Suppose that $p(x)$ is a polynomial function with h zeros and $q(x)$ is a polynomial function of different degree with k zeros. What is the maximum number of zeros that $g(x) = p(x) + q(x)$ could have? What is the maximum number that $f(x) = p(x)q(x)$ could have?   Justify your answers.   (Hint: Look at some special cases and their graphs.)

11. Draw the graph of

$$y = \frac{1}{x - 2}$$

Are there any asymptotes apparent from the graph?   What are they?

12. Draw the graph of

$$y = \frac{x + 4}{x - 1}$$

Are there any asymptotes apparent from the graph?   What are they?

13. Draw the graph of

$$y = \frac{x^2}{x^3 + 1}$$

Are there any asymptotes apparent from the graph?   What are they?

14. Draw the graph of

$$y = \frac{x^2 - 4}{x^2 - 9}$$

with the Standard RANGE setting.   Determine any horizontal or vertical asymptotes.

15. Draw the graph of

$$y = \frac{x^2 + 1}{x^2 - 1}$$

Determine any horizontal or vertical asymptotes.

16. Draw the graph of

$$y = \frac{x^2 - 4}{x - 1}$$

Determine any horizontal or vertical asymptotes.

17. Draw the graph of

$$y = \frac{x - 3}{x^2 - 1}$$

Determine any asymptotes.

18.   Draw the graph of

$$y = \frac{x+2}{x^2+3}$$

Determine any asymptotes.

19.   Draw the graph of

$$y = \frac{x^3 + 4x^2 - 3x - 1}{x^3 - 2x^2 - x + 2}$$

You will need to determine a suitable RANGE setting.   Perhaps more
than one will be needed to complete the exercise.   Determine any
horizontal or vertical asymptotes.

20.   Draw the graph of

$$y = \frac{x^3 - 8}{x^2 - 4}$$

with the RANGE setting [-15,15,2,-20,20,2,1].   How many vertical
asymptotes are there?   Is this the number that you might expect? Does
this conflict with any of the rules of this section?   Explain the result.

21.   Draw the graph of

$$y = \sqrt{x^2 + 3}$$

Are there any asymptotes apparent from the graph?   What are they?

22.   Draw the graph of

$$y = \sqrt{x^2 - 7}$$

Are there any asymptotes apparent from the graph?   What are they?

23.   Draw the graph of

$$y = \frac{6}{\sqrt{x^2 - 4}}$$

Are there any asymptotes apparent from the graph?   What are they?

24.   Draw the graph of

$$y = \frac{3x}{\sqrt{x^2 - 1}}$$

Are there any asymptotes apparent from the graph?   What are they?

25. Draw the graph of

$$y = \frac{x^3 - 8}{x}$$

with a RANGE setting of $[-10,10,1,-20,20,2,1]$. Determine any horizontal or vertical asymptotes. Draw the graph of this function and the function

$$y = x^2$$

with a RANGE of $[-15,-10,5,50,250,50,1]$. Are they similar? How? Explain why.

26. Suppose that the functions $f(x)$ and $g(x)$ have the same horizontal asymptote. What can you say about the function

$$y = f(x) - g(x)$$

Justify your answer.

3

TRIGONOMETRIC FUNCTIONS

Much of nature behaves in periodic or cyclic patterns. Weather cycles, business cycles, and biological clocks are examples of such behavior. The trigonometric functions are periodic and it can be shown that any periodic behavior can be described using trigonometric functions.

The graphics calculator can be used to draw graphs of the six trigonometric functions. Most keyboards only provide three of these but fundamental trigonometric identities enable us to draw the others.

The graphs of the three basic trigonometric functions are given in Figure 3-1. They are drawn with the RANGE setting $[-2\pi,2\pi,\pi/2,-2,2,1,1]$

FIGURE 3-1

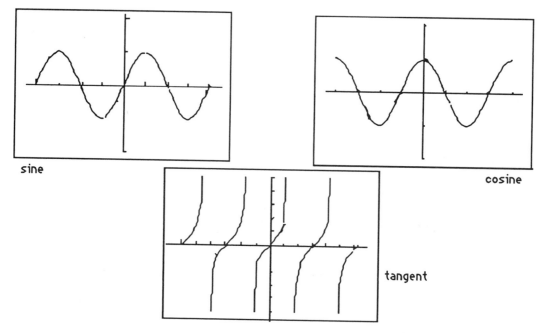

sine

cosine

tangent

The graphs of the cosecant, secant, and cotangent functions can be drawn using the the same RANGE and the identities

$$\csc x = 1/\sin x \qquad \sec x = 1/\cos x \qquad \cot x = 1/\tan x$$

The graphs are given in Figure 3-2.

FIGURE 3-2

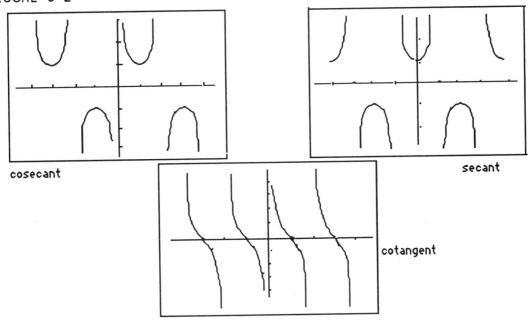

cosecant   secant

cotangent

Graphs of functions involving trigonometric functions can be quite difficult to draw by hand but are quite easy with a graphics calculator.

Example 1. Draw the graph of
$$y = x + \cos(x^2)$$

SOLUTION: The graph is drawn with a RANGE setting of [-5,5,1,-5,5,1,1] .

FIGURE 3-3

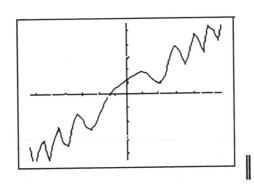

With graphs involving trigonometric functions you may have to try more than one RANGE setting before finding one that serves your purpose.

Since trigonometric functions are periodic, functions that are trigonometric functions of linear functions are also periodic.

Example 2. The function

$$y = \cos(2x + 3) + \sin(3x)$$

is periodic and its period is a multiple of 2π, the period of the sine and cosine functions. The graph is given in Figure 3-4.

FIGURE 3-4

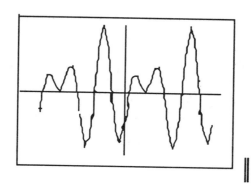

Example 3. Draw the graph of

$$y = 4\sin(2x)$$

and use the graph to estimate the period.

SOLUTION: The graph is given in Figure 3-5; the period appears to be π.

FIGURE 3-5

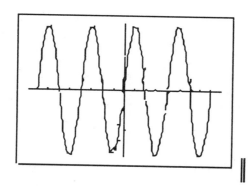

In all the graphs above, x is in radians as is standard practice in calculus. We will follow this convention in this text and advise you to keep your instrument in radian mode. If you have reason to use degrees make sure that your calculator is in degree mode.

Functions such as
$$y = 3\sin(x - 2) + 1$$
are among the most important functions in physics and engineering. The relationship between such a function and the sine function is an example of the shifts discussed in Section 1. The graph of this function is the graph of
$$y = 3\sin x$$
shifted 2 units to the right and 1 upward. The graphs are given in Figure 3-6.

FIGURE 3-6

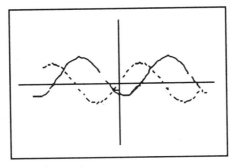

Example 4. Draw the graph of
$$y = \cos(2x + 5) - 3$$

SOLUTION: To see that this is the result of shifting the graph of a simpler function, first rewrite it as
$$y = \cos[2(x + 2.5)] - 3$$
This is the graph of
$$y = \cos(2x)$$
shifted 2.5 to the left and downward 3. Both graphs are given in Figure 3-7.

FIGURE 3-7

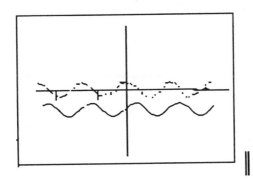

Since trigonometric functions are periodic, they cannot have inverses but they can have partial inverses. In fact, these are so important that they are built into most calculators.

If we restrict the domain of the sine function to $[-\pi/2, \pi/2]$, the partial inverse is denoted by $\arcsin$ or $\sin^{-1}$.

For $-1 \leq y \leq 1$ and $-\pi/2 \leq x \leq \pi/2$:

$\arcsin(y) = x$ or $\sin^{-1}(y) = x$ if and only if $\sin x = y$

If we restrict the domain of the cosine function to $[0, \pi]$, the partial inverse is denoted by $\arccos$ or $\cos^{-1}$.

For $-1 \leq y \leq 1$ and $0 \leq x \leq \pi$:

$\arccos(y) = x$ or $\cos^{-1}(y) = x$ if and only if $\cos x = y$

If we restrict the domain of the tangent function to $(-\pi/2, \pi/2)$, the partial inverse is denoted by $\arctan$ or $\tan^{-1}$.

For any value of y and $-\pi/2 < x < \pi/2$:

$\arctan(y) = x$ or $\tan^{-1}(y) = x$ if and only if $\tan x = y$

The calculator keys for the partial inverse functions are just the seconds or inverses of the keys for the corresponding trigonometric functions. The graphs of these functions are given in Figure 3-8.

FIGURE 3-8

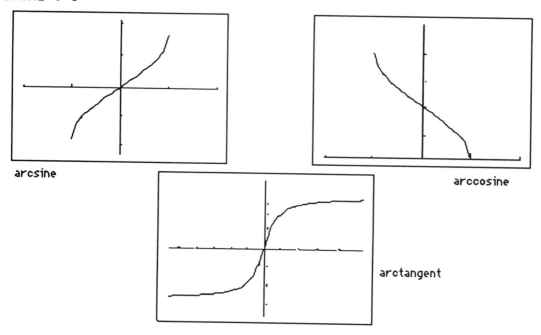

arcsine

arccosine

arctangent

The other inverse trigonometric functions are defined by:

$$\mathrm{arccsc}(t) = \arcsin(1/t) \quad \text{or} \quad \csc^{-1}(t) = \sin^{-1}(1/t)$$

$$\mathrm{arcsec}(t) = \arccos(1/t) \quad \text{or} \quad \sec^{-1}(t) = \cos^{-1}(1/t)$$

$$\mathrm{arccot}(t) = \pi/2 - \arctan(t) \quad \text{or} \quad \cot^{-1}(t) = \pi/2 - \tan^{-1}(t)$$

The graphs of these functions are given in Figure 3-9.

FIGURE 3-9

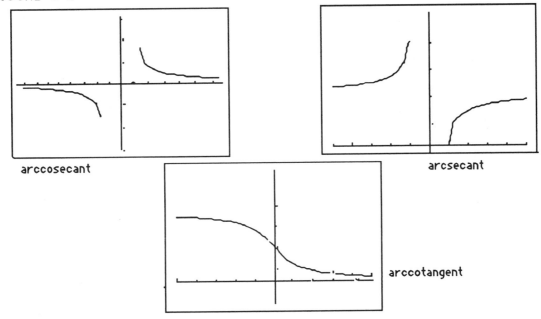

arccosecant   arcsecant

arccotangent

TRIGONOMETRIC IDENTITIES

Trigonometric identities are equations involving trigonometric functions that hold for all meaningful values of the variable. The equations

$$\tan x = \frac{\sin x}{\cos x} \qquad \sec x = \frac{1}{\cos x} \qquad \sin^2 x + \cos^2 x = 1$$

are examples of well known identities.

In order to show that an equation is an identity, we need to transform one side into the other by substitutions or follow an equivalent process. Graphs can be used to help us decide what might be an identity and what is not.

Example 5. Is the equation

$$\frac{1-\sin x}{\cos x} = \frac{\cos x}{1+\sin x}$$

a trigonometric identity?

SOLUTION: We draw the graphs of

$$Y_1 = \frac{1-\sin x}{\cos x} \qquad \text{and} \qquad Y_2 = \frac{\cos x}{1+\sin x}$$

on the same coordinate system in Figure 3-10. They seem to coincide and this indicates that the equation is an identity.

FIGURE 3-10

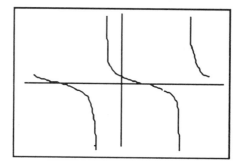

To be sure that we do have an identity, we transform the left side to the right side.

$$\frac{1-\sin x}{\cos x} = \frac{(1-\sin x)(1+\sin x)}{(\cos x)(1+\sin x)} = \frac{1-\sin^2 x}{(\cos x)(1+\sin x)}$$

$$= \frac{\cos^2 x}{(\cos x)(1+\sin x)} = \frac{\cos x}{1+\sin x} \qquad \qquad \|$$

In this text, we will not be concerned with proving identities; we will be content with graphic tests.

Example 6. Determine whether or not the equation

$$(\sin x)(\cos(3x)) = (\sin(2x))(\cos(2x))$$

is an identity.

SOLUTION: We draw the graphs of

$$Y_1 = (\sin x)(\cos(3x)) \qquad \text{and} \qquad Y_2 = (\sin(2x))(\cos(2x))$$

on the same cordinate system in Figure 3-11. Since the graphs are not the same, the equation is not an identity.

FIGURE 3-11

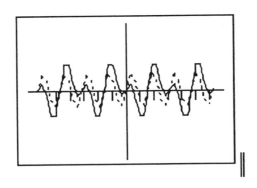

SOLVING TRIGONOMETRIC EQUATIONS

Most equations involving trigonometric functions cannot be solved using algebraic methods but approximate solutions can be found using the techniques from Section 1.

Example 7. Find approximate solutions to the equation
$$\sin^3 x - .2\sin^2 x + .4\sin x + .3 = 0$$

SOLUTION: Let
$$Y_1 = \sin^3 x - .2\sin^2 x + .4\sin x + .3$$
Since the sine function is periodic with period 2π, we draw the graph for $0 \leq x \leq 2\pi$ and look for zeros on that interval. The graph is given in Figure 3-12.

FIGURE 3-12

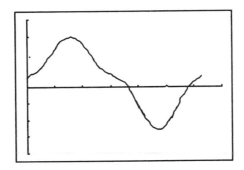

From the graph we see that there is one zero between 3 and 4 and one between 5 and 6.28. We use the **ZERO** program to approximate them. The results are displayed on the screens in Figure 3-13.

FIGURE 3-13

```
?3
ENTER B
?4
ZERO AT
        3.597283733
Y₁=
          3.86E-10
```

```
?5
ENTER B
?6.2B
ZERO AT
        5.82740422B
Y₁=
          2.76E-11
```

We have zeros at x = 3.597283733 and x = 5.827494228.
The rest of the zeros can be obtained by adding even multiples of π to
these.

$$x = 3.597283733 + 2n\pi$$
$$x = 5.827494228 + 2n\pi$$

The trigonometric functions are very important in calculus and its
applications. In real problems, the solutions are not as simple as those found
in textbooks but sufficient solutions can be found using calculators and
computers. As you deal with these functions in calculus, expect to use some of
the techniques developed in this section.

EXERCISES

1. Draw the graphs of $y = \tan x$ and $y = \tan(2x + 3) + 1$

2. Draw the graphs of $y = \sin x$ and $y = 2\sin(3x - 1) + 2$

3. Draw the graph of $y = .5\cos(2x + 1) - 2$

4. Draw the graph of $y = \sin(x^2)$

5. Draw the graph of $y = 2x + \cos x$ for $-20 \leq x \leq 20$

6. Draw the graph of $y = \cos^2 x$

7. Use graphs to decide whether or not

$$\frac{\sin x}{\sec x - \cos x} = \cot x$$

is an identity.

8. Use graphs to decide whether or not
$$\sin^2 x = \frac{1}{2}[1 - \cos(2x)]$$

   is an identity.

9. Use graphs to decide whether or not
$$\sin(3x) = 3\sin x$$

   is an identity.

10. Use graphs to decide whether or not
$$\sin^{-1} x = \pi/2 - \cos^{-1} x$$

   is an identity.

11. Use graphs to decide whether or not
$$\text{arccsc } x = \pi/2 - \text{arcsec } x$$

   is an identity.

12. Approximate any zeros of $y = \sin(2x) + \cos(x+2)$ for $0 \le x \le 2\pi$

13. Approximate any zeros of $y = \tan(\sin(x^2+3))$ for $-2 \le x \le 2$

14. Determine approximate solutions of the equation
$$x^2\sin(3x) = (2x+3)\cos(2x+1) \quad \text{for } -1 \le x \le 1$$

15. Determine any approximate solutions of the equation
$$[\sin(x^2 - 2)]^2 - \cos(3x+1) = x^2$$

16. Draw the graph of $y = \arcsin(\sin(x))$ with the RANGE setting $[-\pi/2, \pi/2, 1, -2, 2, 1, 1]$. Explain why it takes this form.

17. Use the graphs to determine which, if any, of the six basic trigonometric functions are even and which are odd. What about the partial inverses?

18. Determine any vertical or horizontal asymptotes of the six basic trigonometric functions and their inverses. Are any of them related? Explain your answers.

19. Draw the graph of

$$y = \cos\left[\frac{10x}{x^2 + 1}\right]$$

with RANGE setting [0,20,2,-2,2,1,1]. Explain the behavior as x gets larger.

20. Draw the graph of

$$y = \frac{4\sin x}{x}$$

Are there any asymptotes? Explain your answer.

4

LIMITS

The mathematics you studied before calculus was concerned with objects or numbers that were not changing, i.e. you studied fixed figures in geometry and fixed, although sometimes unknown, numbers in algebra. Calculus can be thought of as the mathematics of change. In algebra you may have called a symbol such as x an *unknown*; it is common in calculus to call it a *variable* because you are concerned with how other quantities change as x changes.

The concept of a *limit* lies at the core of calculus. We give both the intuitive notion and the formal definition.

Let L be a real number, f a function that is defined (except possibly at c) on an open interval containing c:

INTUITIVE NOTION: $\lim_{x \to c} f(x) = L$ means that f(x) becomes arbitrarily close to L as x approaches c.

FORMAL DEFINITION: $\lim_{x \to c} f(x) = L$ means that for each $\varepsilon > 0$ there exists a $\delta > 0$ such that

$|f(x) - L| < \varepsilon$ whenever $0 < |x - c| < \delta$

The formal definition is illustrated in Figure 4-1.

FIGURE 4-1

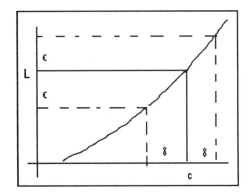

The graphics calculator can be used to better understand the formal definition.

Example 1. Determine $\lim_{x \to 0} \left(\dfrac{\sin x}{x} \right)$.

SOLUTION: The graph is given in Figure 4-2 with RANGE setting [-2,2,1,-.5,1.5,1,1].

FIGURE 4-2

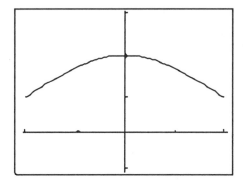

It appears that y is close to 1 for x close to 0 but we know that no value of this function exists at x = 0 because division by 0 is undefined. Suppose we let ε = .1, set the y values in the RANGE to .9 and 1.1, but do not change the x values. The graph is given in Figure 4-3.

FIGURE 4-3

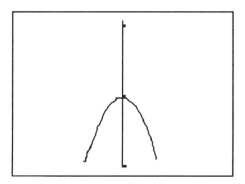

In this case, only part of the curve is visable; in fact only the part where .9 ≤ y ≤ 1.1. If we use TRACE and shift the cursor left and right, we see that $\left| y - 1 \right| < .1$ for -.77894737 < x < .7789737; otherwise the y values are more than .1 away from 1. If we let δ = .77, then $\left| y - 1 \right| < .1$ whenever $0 < \left| x - 0 \right| < .77$.

If we let ε = .01, set the y values from .99 to 1.01, and set the x values from -.77 to .77, we again get only a portion of the curve as you see in Figure 4-4.

FIGURE 4-4

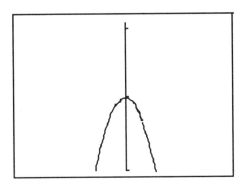

This time as we move the cursor left and right, we see that
$$|y - 1| < .01 \quad \text{whenever} \quad 0 < |x - 0| < .23.$$
In this case, then, a δ value of .23 will suffice for an ε value of .01. We could repeat the process with ε = .001 and so on and in each case obtain a suitable δ value. It appears that for each value of ε > 0 there is a value for δ such that
$$|y - 1| < \varepsilon \quad \text{whenever} \quad 0 < |x - 0| < \delta$$
and we would expect that $\lim\limits_{x \to 0} \left(\dfrac{\sin x}{x} \right) = 1$. We have not proved that the statement holds but we have illustrated it graphically. ‖

The expressions
$$\lim\limits_{x \to c^+} f(x) = L \qquad \text{and} \qquad \lim\limits_{x \to c^-} f(x) = L$$

denote the limits from the right and left at c. The definitions are similar to those given before and we refer you to your calculus text.

LIMITS AT INFINITY

The expression
$$\lim\limits_{x \to \infty} f(x) = L$$

can be interpreted intuitively as
"f(x) gets close to L as x gets arbitrarily large."

The formal definition is similar to the earlier one.

FORMAL DEFINITION: $\lim\limits_{x \to \infty} f(x) = L$ if for each $\varepsilon > 0$, there exists an

M > 0 such that $\left| f(x) - L \right| < \varepsilon$ whenever x > M.

The definition of $\lim\limits_{x \to -\infty} f(x)$ is similar.

The intuitive idea of the limit as x→∞ is closely related to the concept of a horizontal asymptote: as x gets large, f(x) gets close to the line y = L.

Example 2. If we draw the graph of

$$y = 3 + \frac{2}{x}$$

with RANGE setting [-10,10,2,-1,7,1,1], we see that the curve gets close to the line y = 3. This is illustrated in Figure 4-5 with

$Y_1 = 3 + \dfrac{2}{x}$ and $Y_2 = 3$.

FIGURE 4-5

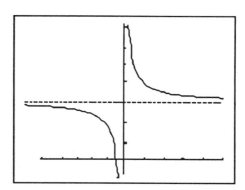

If we let ε = .1 and change the y settings to 2.9, 3.1, there are no points on the curve near the line but if we change the x values to 10 and 1000, part of the curve appears, the part with y values between 2.9 and 3.1 is shown in Figure 4-6.

FIGURE 4-6

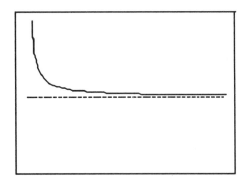

If we use TRACE, we see that any value of x greater than 20 yields a y value within .1 of 3. Thus it appears that

$$|y - 3| < .1 \text{ whenever } 20 < x.$$

Similarly, if ε = .01 and M = 1000,

$$|y - 3| < .01 \text{ whenever } 1000 < x.$$

This process can be continued as before and illustrates that

$$\lim_{x \to \infty} \left(3 + \frac{2}{x}\right) = 3$$

COMPUTING LIMITS

How do we determine limits? The formal definitions help us to understand limits and can be used to verify that certain limit statements are true but they do not provide any computational techniques. Your calculator may have such an approximation process but if not, the programming capacity of your calculator can be used to approximate limits.

The program below is written for the TI-81 but can be adapted to most programmable calculators. The values of the function are computed for 12 values of x above c and 12 values of x below c. The successive differences between c and the numbers where the values of the function are computed are .1, .01, .001, $\cdots$, .0000000000001. If these values do not suffice, then the computation is beyond the capacity of the calculator and the program stops. The program also approximates limits from the right and left and tests to see if they are the same.

LIMIT

```
Prgm :LIMIT
:10000→P
:-10000→Q
:0→N
:.000000001→E
:Disp"ENTER C"
:Input C
:.1→D
:Lbl 1
:C+D→X
:If abs(Y₁-P)<E
:Goto 6
:Y₁→P
:C-D→X
:If abs(Y₁-Q)<E
:Goto 7
:Y₁→Q
:D*.1→D
:1+N→N
:If N>12
:Goto 5
:Goto 1
:Lbl 5
:Disp "AFTER 12
APPROXIMATIONS N
O LIMIT"
:Goto 9
```

```
:Lbl 6
:Disp"RIGHT LIM
IT"
:Disp P
:Lbl 7
:Disp "LEFT LIM
IT"
:Disp Q
:If abs(P-Q)>10
0*E
:Goto 4
:(P+Q)/2→W
:Disp"LIMIT IS"
:Disp W
:Goto 9
:Lbl 4
:Disp"RIGHT AND
LEFT LIMITS DI
FFER OR ONE DOES
NOT EXIST, NO L
IMIT"
:Lbl 9
:End
```

Example 3. Approximate

$$\lim_{x \to 3} \left(\frac{x^3 - 6x^2 + 2x + 21}{x^2 - 9} \right)$$

SOLUTION: We use the LIMIT program with the function as Y_1 and C = 3. The results are given by the screen in Figure 4-7

FIGURE 4-7

```
?3
RIGHT LIMIT
          -1.166666667
LEFT LIMIT
          -1.166666667
LIMIT IS
          -1.166666667
```

The limit is about -7/6. ‖

The LIMIT program also shows you when the right and left limits exist and are different.

Example 4. Compute the right and left limits of

$$y = \frac{x}{|x|} \text{ at } c = 0$$

SOLUTION: The graph is given in Figure 4-8.

FIGURE 4-8

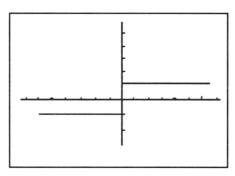

The results of the LIMIT program are displayed in Figure 4-9.

FIGURE 4-9

```
RIGHT LIMIT
                        1
LEFT LIMIT
                       -1
RIGHT AND LEFT
LIMITS DIFFER OR
ONE DOES NOT EX
IST, NO LIMIT
```

Both one-sided limits exist but are not the same. ‖

Limits at infinity or negative infinity can be approximated using the **LIMIT** program and the fact that:

$$\lim_{x \to \infty} f(x) = \lim_{x \to 0^+} f(x^{-1}) \quad \text{and} \quad \lim_{x \to -\infty} f(x) = \lim_{x \to 0^-} f(x^{-1})$$

We just need to replace X by X^{-1} in the function and apply the **LIMIT** program.

Example 5. Approximate

$$\lim_{x \to \infty} \frac{2x^2 + \sin x}{3x^2 + \cos x}$$

SOLUTION: Let

$$Y_1 = \frac{2(X^{-1})^2 + \sin(X^{-1})}{3(X^{-1})^2 + \cos(X^{-1})}$$

and let C = 0. We need to determine the limit from the right; the results are displayed in Figure 4-10.

FIGURE 4-10

```
?0
RIGHT LIMIT
        .6666666667
LEFT LIMIT
        .6666666667
LIMIT IG
        .6666666667
```

The limit from the right is about 2/3 and, therefore, $\lim_{x \to \infty} Y_1 = 2/3$.

Notice that the limit from the left is also about 2/3 and, therefore, $\lim_{x \to -\infty} Y_1$ is also about 2/3. ‖

Remember that your calculator can only provide approximations. These suffice for most functions but you must be on guard for the unusual cases where the approximations by the calculator are not adequate.

Example 6. Approximate

$$\lim_{x \to \infty} \frac{x^x}{10000^x}$$

SOLUTION: If we attempt to approximate this using the program, we get an error message because the numbers being computed get too large. On the other hand, we can see that for $x > 10000$, the fraction is greater than 1 and raising to the x power gives arbitrarily large numbers. ▊

Expressions such as

$$\lim_{x \to c^+} f(x) = \infty$$

and others obtained by suitable sign changes can be interpreted as saying that $f(x)$ gets large (or large in absolute value but negative) as x approaches c from one side or the other. The graphics calculator can be used to estimate many such points by drawing the graph of the function and then using TRACE. The graph will have a vertical asymptote at these points and your calculator will display it for you if the RANGE setting is suitable.

Example 7. Approximate any points c where one of the one sided limits at c is either ∞ or $-\infty$ for the function

$$y = \frac{4x}{x - 3}$$

SOLUTION: The graph is given in Figure 4-11.

FIGURE 4-11

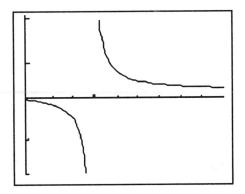

If we use TRACE, we see that as values of x are chosen close to 3 on the left, the y values get large in absolute value but are negative.

As we choose values of x close to 3 but on the right, the y values become arbitrarily large.

From these graphic results we would expect that
$$\lim_{x\to3^-} \frac{4x}{x-3} = -\infty \quad \text{and} \quad \lim_{x\to3^+} \frac{4x}{x-3} = \infty$$

The graphic results do not prove the conclusions but they certainly lead us in the right direction. ‖

In the case of more complicated functions, the techniques we used for approximating vertical asymptotes in Section 2 will provide the information.

Example 8. Approximate any points c where one of the one sided limits is either ∞ or -∞ if
$$y = \frac{5}{x^3 + 4x - 7}$$

SOLUTION: The graph is given in Figure 4-12

FIGURE 4-12

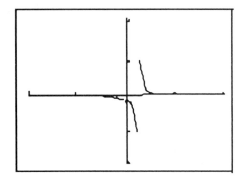 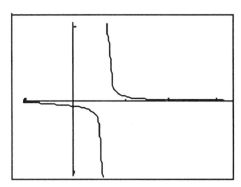

The first drawing uses the Standard RANGE; the second is [-2,6,2,-5,5,1,1].   The curve jumps from negative values of y to positive values of y as x changes from 1 to 2. If we list the denominator as Y_1 and apply the ZERO program, we can approximate the x coordinate of the vertical asymptote.   The results are displayed in Figure 4-13.

FIGURE 4-13

```
?1
ENTER B
?2
ZERO AT
        1.255383157
Y₁=
            -1.176E-9
```

Combining this information with Figure 4-12, we see that the function approaches -∞ from the left and ∞ from the right for c about 1.255. ‖

The techniques given here are no substitute for general techniques used in computing limits; they only provide estimates of the numbers we need to find. There are many methods in calculus for determining limits and these should be in your collection of skills.

EXERCISES

1. Use a graph to estimate $\lim_{x \to 2} \left(\dfrac{x^5 - 32}{x^2 - 4} \right)$. Find a value of δ that suffices for $\epsilon = .1$. Find a δ for $\epsilon = .01$.

2. Use the graph to estimate $\lim_{x \to 1} \left(\dfrac{\sin(x^3 - 1)}{x^3 - 1} \right)$. Find a value of δ that suffices for $\epsilon = .1$. Find a δ for $\epsilon = .01$.

3. Use **LIMIT** to approximate $\lim_{x \to 3} \left(\dfrac{x^2 - 9}{x^3 - 27} \right)$.

4. Use **LIMIT** to approximate $\lim_{x \to 0} \left(\dfrac{1 - \cos x}{x} \right)$.

5. Use **LIMIT** to approximate $\lim_{x \to 0} \left(\dfrac{\sin(3\tan x)}{x} \right)$.

6.   Use **LIMIT** to approximate $\displaystyle \lim_{x \to 0} \left(\frac{[1/(x+5) - (1/5)]}{x} \right)$.

7.   Use **LIMIT** to approximate $\displaystyle \lim_{x \to 1} \frac{\sin(x^2 - 1)}{x - 1}$

8.   Use **LIMIT** to approximate $\displaystyle \lim_{x \to 3} \ \sin\left(\frac{x^3 - x^2 - 18}{x^2 - 14x + 33} \right)$.

9.   Use **LIMIT** to approximate $\displaystyle \lim_{x \to -2} \ \cos\left(\frac{x^3 + 8}{x^5 + 32} \right)$. Let $\varepsilon = .1$ and find a sufficient δ.

10.   Use **LIMIT** to aproximate $\displaystyle \lim_{x \to 1} \left(\frac{\sin(x-1)}{\cos(x-1)} \right)$. Let $\varepsilon = .1$ and find a sufficient δ.

11.   Use **LIMIT** to approximate $\displaystyle \lim_{x \to 2^+} \left(\frac{x^2 - 4}{|x - 2|} \right)$.

12.   Use **LIMIT** to approximate $\displaystyle \lim_{x \to 3^-} \left(\frac{|x^2 - 9|}{x - 3} \right)$. Let $\varepsilon = .1$ and find a sufficient δ.

13.   Use **LIMIT** to approximate $\displaystyle \lim_{x \to \infty} \left(\frac{\sin x}{x + 1} \right)$.

14.   Use **LIMIT** to approximate $\displaystyle \lim_{x \to -\infty} \left(\frac{2x^3 - 7x^2 + 3x - 2}{5x^3 + 4x^2 + 10x - 5} \right)$.

15.   Use **LIMIT** to approximate $\displaystyle \lim_{x \to -\infty} \left(\frac{\sin^3(2x + 3)}{x^{.7}} \right)$.

16.   Use **LIMIT** to approximate $\displaystyle \lim_{x \to \infty} \left(\frac{\sqrt{5x^4 + 3x^2 + 19}}{x^2 + 5} \right)$. Let $\varepsilon = .1$ and find a sufficient M. (Hint: Find a sufficient δ as $x^{-1} \to 0^+$ and let $M = 1/\delta$.)

17. Determine any numbers c where the limit from the right or from the left is infinite for the function $f(x) = \dfrac{x^2 + 5x + 9}{x^2 + 8x - 2}$.

18. Determine any numbers c where the limit from the right or from the left is infinite for the function $f(x) = \dfrac{\sin(x^2 + 7)}{x^3 + 4x^2 - 7x - 1}$.

19. Find a function $f(x)$ such that $\lim\limits_{x \to 4} |f(x)|$ exists but $\lim\limits_{x \to 4} f(x)$ does not.

Explain your result.

20. Explain why $\lim\limits_{x \to c} \dfrac{\sin x}{g(x)} = 0$ if $\lim\limits_{x \to \infty} g(x) = \infty$. (Hint: Construct an example and study the graph.)

5

DERIVATIVES

Calculus can be viewed as the mathematics of change and derivatives as rates of change. The derivative of a function $f(x)$ is denoted by $f'(x)$ and defined by

$$f'(x) = \lim_{\Delta x \to 0} \frac{f(x + \Delta x) - f(x)}{\Delta x}$$

providing the limit exists.

We can get a good grasp of the concept by thinking of the derivative of a function at a point as the slope of the tangent line to the graph at the point.

Example 1. Draw the graph of
$$f(x) = .5x^3$$
and then draw the graphs of the secant line through the points $(2,4)$ and $(3,13.5)$, the secant line through the points $(2,4)$ and $(2.5,7.8125)$, and the tangent line at the point $(2,4)$.

SOLUTION: The slopes of the two secant lines are 9.5 and 7.625 and the slope of the tangent line is 6.   The   equations of the lines are
$$y - 4 = 9.5(x - 2)$$
$$y - 4 = 7.625(x - 2)$$
$$y - 4 = 6(x - 2)$$

We draw the four graphs on one coordinate system by letting
$$Y_1 = .5X^3$$
$$Y_2 = 9.5(X - 2) + 4$$
$$Y_3 = 7.625(X - 2) + 4$$
$$Y_4 = 6(X - 2) + 4$$

The graphs are given in Figure 5-1 with a **RANGE** of   [1,3,1,0,15,1,1]

FIGURE 5-1

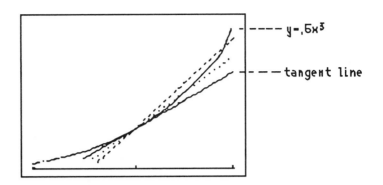

The secant lines are very close to the tangent line. This illustrates the idea that the slope of the tangent line is the limit of the slopes of the secant lines. ‖

Example 2. Draw the graphs of
$$y = x^2 \sin x$$
and the tangent line to the curve at the point where $x = 2.5$.

SOLUTION: First we let
$$Y_1 = y = X^2 \sin X$$
$$Y_2 = y' = X^2 \cos X + 2X \sin X$$
When $x = 2.5$, we need to compute Y_1 to determine the y coordinate of the point and Y_2 to determine the slope of the line. As we compute these values, we can store them in A and B to eliminate copying later. The screens appear as in Figure 5-2.

FIGURE 5-2

```
2.5→K
              2.5
Y₁
         3.740450901
Ans→A
         3.740450901
```

```
Y₂
         -2.014786877
Ans→B
         -2.014786877
```

Now let
$$Y_3 = B(X - 2.5) + A,$$
clear Y_2, and draw. The graphs are given in Figure 5-3.

67

FIGURE 5-3

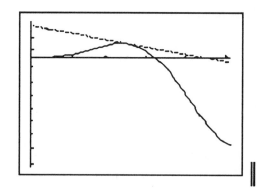

THE MEAN VALUE THEOREM

Many properties of derivatives can be illustrated using a graphics calculator; the following theorem is among them.

The Mean Value Theorem for Derivatives:
If f is continuous on [a,b] and differentiable on (a,b), then there is a number c in (a,b) such that
$$f'(c) = \frac{f(b) - f(a)}{b - a} \ .$$

Example 3. Let $f(x) = x^2 + \cos^2 x$ on [1,3].

(a) Find a number c such that
$$f'(c) = \frac{f(3) - f(1)}{3 - 1}$$

Draw the following on one coordinate system:
(b) The graph of f(x) on an interval containing [1,3].
(c) The secant line through the points (1,f(1)) and (3,f(3)).
(d) The line tangent to the graph of y = f(x) at the point where x = c.

SOLUTION: First let
$$Y_2 = y = X^2 + (\cos X)^2$$
and
$$Y_1 = y' = 2X - 2\cos X \sin X$$

As before, we will store the computed values to simplify future entries. The results are displayed in Figure 5-4.

FIGURE 5-4

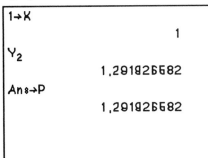

1→K	
	1
Y₂	
	1.291926582
Ans→P	
	1.291926582

3→K	
	3
Y₂	
	0.980085143
Ans→Q	
	0.980085143
(Q-P)/2→W	
	4.344079281

We have computed $f(1)$, $f(3)$, and $\dfrac{f(3)-f(1)}{3-1}$ stored them in P, Q, and W. Now the problem is to find c such that $f'(c) = W$.
Solving the equation

$$f'(c) = W$$

reduces to solving the equation

$$f'(c) - W = 0$$

If we modify Y_1 by subtracting W and find a zero of Y_1, we will find the desired value for c. The Mean Value Theorem assures us that there is a solution between 1 and 3 and the **ZERO** program can be used to approximate it. The results are given in Figure 5-5.

FIGURE 5-5

?3	
ZERO AT	
	1.881196479
Y₁=	
	1.57E-9
Y₂=	
	3.632193596

X→C	
	1.881196479
Y₂→D	
	3.632193596

The value of c is approximated by 1.881196479 and we store it in C. The value of $f(c)$ is approximated by 3.632193596 and we store it in D. Now clear Y_1 and let

$$Y_3 = W(X - 3) + Q$$
$$Y_4 = W(X - C) + D$$

The graphs are drawn in Figure 5-6 and, as expected, the tangent line is parallel to the secant line.

FIGURE 5-6

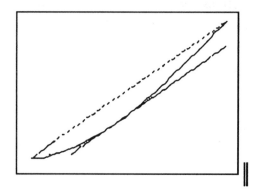

Tangent lines provide good approximations of curves for small intervals.

Example 4. How does the graph of the tangent line at (1,1) approximate the graph of

$$y = x^3?$$

SOLUTION: The graphs of the curve and the tangent line at (1,1) are given in Figure 5-7 with RANGE [0,2,1,0,10,5,1].

FIGURE 5-7

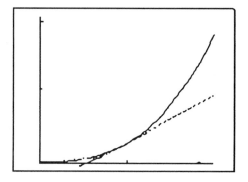

The curve and the line seem to coincide from approximately x = .8 to x = 1.2. They are not the same but appear to be because the distance between the curve and the line is less than the width of the graph. ‖

Other properties of derivatives can be illustrated using the graphics calculator and we include some of these in the exercises. Numerical approximations of derivatives are sufficiently important that they are treated by themselves in the next section.

EXERCISES

1.　Let $f(x) = x^3 + 2x - 4$. Draw the graphs of $f(x)$, the secant line through the points where $x = 2$ and $x = 2.5$, and the tangent line at the point where $x = 2$.

2.　Let $f(x) = \dfrac{4x}{x^2+2}$. Draw the graphs of $f(x)$, the secant line through the points where $x = 1$ and $x = 1.3$, and the tangent line at the point where $x = 1$.

3.　Let $f(x) = \sin x$. Draw the graphs of $f(x)$, the secant line through the points where $x = 2$ and $x = 2.2$, and the tangent line at the point where $x = 2.2$.

4.　Let $f(x) = x^3 \cos x$. Draw the graph of $f(x)$, the secant line through the points where $x = 3$ and $x = 3.3$, and the tangent line at the point where $x = 3$.

5.　Let $f(x) = \dfrac{\cos x}{3+\sin x}$. Draw the graph of $f(x)$ and the tangent line at the point where $x = 2$.

6.　Let $f(x) = \tan(x^2)$. Draw the graph of $f(x)$ and the tangent line at the point where $x = 1$.

7.　Let $f(x) = (x^4 - 10)^2$. Draw the graph of $f(x)$ and the tangent line at the point where $x = 2$.

8.　Let $f(x) = \sin(\cos(x^2))$. Draw the graph of $f(x)$ and the tangent line at the point where $x = 2$.

9.   Let $f(x) = .5x^4 + x^2$ on $[1,2]$. Find a value of c in $(1,2)$ such that
     $f'(c) = f(2) - f(1)$

10. Let $f(x) = \sin^3(x^2)$ on $[1,3]$. Find a value of c in $(1,3)$ such that
     $$f'(c) = \frac{f(3) - f(1)}{3-1}$$

11. Let $f(x) = x^3 + 3x - 7$.

     (a) Find a value of c such that $f'(c) = \dfrac{f(5) - f(2)}{5-2}$

     (b) Draw the graphs of $f(x)$ and the secant line through $(2,f(2))$ and $(5,f(5))$.

     (c) Draw the graph of the tangent line at the point where $x = c$.

12. Let $f(x) = (\cos(3x+1) + 2)^3$.

     (a) Find a value of c such that $f'(c) = \dfrac{f(4) - f(1)}{4-1}$.

     (b) Draw the graphs of $f(x)$ and the secant line through $(1,f(1))$ and $(4,f(4))$.

     (c) Draw the graph of the tangent line at the point where $x = c$.

13. Draw the graph of $y = x^{2/3} + x$ and the graph of the line tangent to the curve at the point where $x = 8$. Give an interval upon which the curve and the line seem to coincide.

14. Draw the graph of $y = \sin x$ and the line tangent to the curve at the point where $x = .5$. Give an interval upon which the curve and the line seem to coincide.

15. Let $g(x) = x^2 \sin x$. Find a number c where $g'(c) = .1$.

16. Let $g(x) = \cos(\sin x)$. Find a number where $g'(c) = -.2$.

EXTENSION

17.   Let $p(x) = x^3 + 4x - 5$ and $q(x) = x^4 + 3x^2 - 4$ and $f(x) = \dfrac{p(x)}{q(x)}$.

(a)   Use LIMIT to approximate $\displaystyle\lim_{x \to 1} f(x)$.

(b)   Compute $\dfrac{p'(1)}{q'(1)}$

(c)   Compare the results of (a) and (b).   What can you say?

18.   Let p and q be as in Exercise 17.   Find a number c such that
$$\frac{p'(c)}{q'(c)} = \frac{p(3) - p(1)}{q(3) - q(1)}$$

6

NUMERICAL DIFFERENTIATION

At times you may wish to know an approximation of the derivative of a function at a particular value without computing the entire derivative and then evaluating. For example, you may wish to determine if a function is increasing at a point and it will suffice to know that the derivative is positive at the point. Such an approximation can be computed with any calculator using the difference quotient but some calculators have a built-in process for doing so.

The derivative

$$f'(x) = \lim_{\Delta x \to 0} \frac{f(x + \Delta x) - f(x)}{\Delta x}$$

can be approximated by the quotient

$$\frac{f(x + \Delta x) - f(x - \Delta x)}{2 \Delta x}$$

for small values of Δx. This expression usually yields a better aproximation than is obtained by just choosing small values of Δx and evaluating the difference quotient given in the definition of the derivative. A geometric illustration is given in Figure 6-1.

FIGURE 6-1

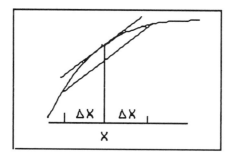

THE PROCESS FOR THE TI-81:

1. Store the value of x for which the approximation is to be computed.

2. Activate NDeriv by using MATH 8

3. List the function in terms of X and follow this by a comma.

4. List your choice of Δx and press ENTER.

5. The approximation is displayed.

In most cases, the smaller Δx, the better the approximation.

Example 1. The force of gravitation between two bodies is inversely proportional to the square of the distance between them. This means that

$$F = \frac{k}{r^2}$$

where r is the distance between them.

If k = 1000, estimate the rate of change of F when r = 10.

SOLUTION: We can apply the NDeriv process to estimate the derivative of F at 10. In this case we will let r = X and let X change by .01.

First store 10 in X.

From the MATH menu choose 8:NDeriv

The screen now appears as in Figure 6-2.

FIGURE 6-2

```
10→X
                        10

NDeriv(
```

Now enter the function, a comma, and the value for ΔX. In this case let Δx = .01. After pressing ENTER, the screen will appear as in Figure 6-3.

FIGURE 6-3

```
10→X
                   1 0
NDeriv(100D/X²,.
01)
           -2.0D0004
```

The derivative is approximated by -2.000004.

If we solve the problem in the usual way, we find that $\dfrac{dF}{dr} = -2$ at $r = 10$. The numerical approximation was very good. ‖

The approximation process is most useful in cases where the function is quite complicated and computing the derivative would be very time consuming.

Example 2. Approximate the derivative of

$$y = \sqrt[5]{\dfrac{x^2 + 8x + 9}{x^3 - 7x^2 + 3x + 12}}$$

at $x = .3$.

SOLUTION: First we store .3 in X.

From the MATH menu, choose 8:NDeriv., enter the function and the value for ΔX. In this case let $\Delta X = .0001$. The results are displayed in Figure 6-4.

FIGURE 6-4

```
.3→X
                          .3
NDeriv(((x²+8x+9
)/(x³-7x²+3x+12
)^.2,.0001)
            .1625985955
```

The derivative at .3 is approximately .1625985955. ‖

Example 3. Suppose that experience and statistical analysis has shown that the profits in your business are approximated by
$$y = 1245\sin^2(.00001x^2)$$
where x is the number of units produced. If you are producing 1000 units, is your profit increasing or decreasing and at what rate?

SOLUTION: To answer this question we need to know whether the derivative of the profit function at x = 1000 is positive or negative. The screen in Figure 6-5 gives the approximation with $\Delta X = 1$.

FIGURE 6-5

```
1000→X
                1000
NDeriv(1245(sin
(.00001x2)2,1)
        22.72647844
```

The derivative at x=1000 is positive and, therefore, at a production level of 1000 units, profits are increasing at a rate of about $22.73 per unit as production increases. ‖

Example 4. Suppose that the velocity of an object at time t is given by

$$V = \frac{10\cos^2(3x^4+2)}{x^2+2}$$

Approximate the acceleration at time x = 3.5.

SOLUTION: We know that the acceleration is the derivative of the velocity and we use the numerical differentiation process to approximate the acceleration at time x = 3.5. We choose $\Delta x = .01$. The results are displayed on the screen in Figure 6-6.

FIGURE 6-6

```
3.5→X
                3.5
NDeriv(((10(cos
(3X4+2)l2)/(x2
+2)),.01)
        -9.46748458
```

The acceleration is about -9.47 at time x = 3.5. ‖

The numerical differentiation process can be combined with the LIMIT program from Section 4 to provide even better approximations for some derivatives. Just list Y_1 as NDeriv(Y_2,D) and Y_2 as the function. The limit is the approximation of the derivative. In this case the right and left limits are the same.

Example 5. Approximate f'(.4) if
$$f(x) = (\sin(2x+3))^3 + \tan(5x-1)$$

SOLUTION: First enter
$$Y_1 = \text{NDeriv}(Y_2,D)$$
$$Y_2 = (\sin(2x+3))^3 + \tan(5x-1)$$

Now use the LIMIT program entering .4 as A. The results are displayed on the screen in Figure 6-7.

FIGURE 6-7

```
?.4
RIGHT LIMIT IS
            15.35
LEFT LIMIT IS
            15.35
LIMIT IS
            15.35
```

The derivative is approximately 15.35. ‖

The limit program does not always provide the desired approximation. The rounding errors resulting from the division when using NDeriv may keep the approximations from getting close enough to look like a limit. This does not mean that numerical differentiation is not valuable; it just means that we cannot expect as high degree of precision as we might hope for from this approximation process.

If your calculator does not have a built-in process for computing numerical approximations of derivatives, you can let
$$Y_1 = \frac{f(X) - f(C)}{X - C}$$

and use LIMIT to approximate the value of the limit of this quotient as X approaches C.

Numerical approximation of derivatives is useful any time you have a complicated derivative to compute and only need an approximation of the value of the derivative at one point. It does not provide you with the general forms that are needed to solve many of the problems that arise in calculus.

EXERCISES

1. Compute an approximation of $f'(3)$ with $\Delta x = .01$ if
$$y = f(x) = \frac{x^3 \sin(4x+3)}{x^3 + 5x + 3}$$

2. Compute an approximation of $f'(2)$ with $\Delta x = .001$ if
$$y = f(x) = \cos(x^2 + 4x^3 + 5x^{-3})$$

3. Compute an approximation of $g'(5)$ with $\Delta x = .01$ if
$$y = g(x) = \sqrt[3]{\cos(x^2 + x + 5) + \sin^2(x^{.6})}$$

4. Compute an approximation of $g'(4)$ with $\Delta x = .1$ if
$$y = g(x) = \sqrt{\frac{x^{.7} \cos^2(x^{1.2} + 3x^{.4})}{x^{.6} + 5}}$$

5. Compute an approximation of $p'(.4)$ with $\Delta x = .1$ if
$$y = p(x) = \tan\left[\frac{\sin^4(3x)}{x^2 + 4} + x^3\right] + 7$$

6. Compute an approximation of $p'(2)$ with $\Delta x = .1$ if
$$y = p(x) = \sin(\cos(\tan(x)))$$

7. Compute an approximation of $f'(1.3)$ with $\Delta x = .001$ if
$$y = f(x) = \sec\left[x^3 + 4x^2 + 2x + 1\right]$$

8. Compute an approximation of $g'(3)$ with $\Delta x = .001$ if
$$y = g(x) = \left[\cos^6(x^2+1)+x^3+5\right]^{.19}$$

9. Compute an approximation of $q'(2)$ with $\Delta x = .001$ if
$$y = q(x) = \left|\sin(x^2 + 3) - \cos(x^3-7)\right|$$

10. Use **LIMIT** to approximate $f'(.3)$ if $f(x) = x\sin(x^2)$. Use calculus to check your result.

11. Use **LIMIT** to approximate $g'(2.4)$ if $g(x) = \sin(\cos x)$

12. If an object is moving in a linear path with position function
$$s = \sqrt{\tan(2x)+x^3-2x}$$
approximate its velocity when $x = 1.7$.

13. Write an equation that is an estimate of the tangent line to the graph of
$$y = \cos(\sin(x))$$
at the point where $x = 1.4$.

14. If the profits of a business are approximated by
$$p = \sqrt[3]{\sin^2(.001x^2)+500x+200x^{-3}}$$
approximate the rate of change of the profits at a production level of 100 units. (Since you cannot produce a fractional part of a unit, you can let $\Delta x = 1$.)

15. Suppose the height in feet of an object above the ground is given by
$$h = \sqrt{x^2+\sin(4x+3)+\cos(2x-1)+3}$$
at time x measured in seconds.. Is the object moving upward or downward at time $x = 3$? At what rate?

7

MAXIMA AND MINIMA

The techniques developed in calculus for finding maximum and minimum values of functions depend upon our ability to determine the values of the variable where the derivative of the function is zero. In many cases this is very difficult or nearly impossible. Approximate solutions to many maximum and minimum problems can be found using a graphics calculator. In this section two methods are demonstrated along with applications of the techniques.

GRAPHICS APPROXIMATION METHOD:

This method uses graphs of functions along with TRACE and ZOOM.

Example 1: Determine any local or relative[1] maxima and minima for
$$f(x) = x^4 - 5x^3 + 4x^2 + 6x + 5$$

SOLUTION: The graph is given twice in Figure 7-1. The Standard RANGE setting of [-10,10,1,-10,10,1,1] yields the first drawing and [-3,5,1,-20,20,5,1] gives the second. The second graph provides a better view of the behavior of the function.

FIGURE 7-1

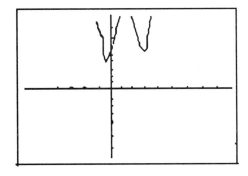 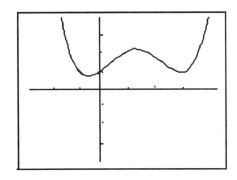

We see that the graph changes direction at least three times with at least two relative minima and one relative maximum. We can estimate the x and y values at these points using the TRACE and ZOOM keys.

1.The words "local extrema" and "relative extrema" have the same meaning..

From the first graph in Figure 7-2, we see that, for the first minimum, the original estimates are x = -.3894737 and y = 3.5883225.

If we use ZOOM IN with the ZOOM factors set at 4, the second graph gives a better estimate of x =-.4 and y = 3.5856.

Another application of ZOOM IN yields x=-.4078947 and y = 3.5851495. A third application gives x= .4059211 and y = 3.585133 and a fourth yields x=-.4064145 with y = 3.585129.

Another application does not improve the accuracy; we suggest you try it and compare the results of your last two approximations. (Remember the discussion of accuracy in Section 1.)

FIGURE 7-2

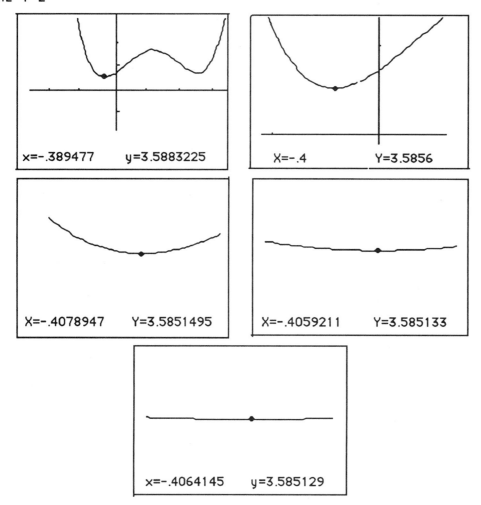

The other extreme values can be found in a similar manner and we leave doing so as an exercise. ‖

COMPUTATIONAL APPROXIMATION METHOD:

This method uses the graphs of the function and its derivative along with the **ZERO** program or a similar process.

A slight editing of the **ZERO** program from Section 1 will produce the maximum and minimum values of y at the same time that it determines the corresponding values of x. Just add $'Disp"Y_2="'$ and $'Disp\ Y_2'$. With this method we list y' as Y_1 and y as Y_2 since we need to find the points where $y' = 0$.

Example 2: Find any local or relative extrema for
$$y = \frac{5 + \sin x}{2 + \cos x}$$

SOLUTION: We know that the trigonometric functions are periodic with period 2π and, therefore, we only need to study the graph on $[0,2\pi]$ to determine all of the behavior of the function. Since the numerator must be between -4 and 6 and the denominator is between 1 and 3, a suitable **RANGE** setting for y will be from -4 to 6. We use the setting $[0,2\pi,\pi/2,-4,6,1,1]$. The graph is given in Figure 7-3 letting Y_2 be the function.

FIGURE 7-3

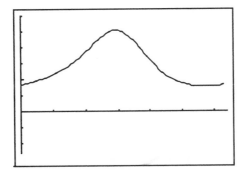

The curve appears to have a local minimum between 4 and 6 with a local maximum between 2 and 4 .

The derivative is

$$y' = \frac{2\cos x + 5\sin x + 1}{(2 + \cos x)^2}$$

Since the denominator of this fraction is always a positive number, $y' = 0$ exactly when

$$2\cos x + 5\sin x + 1 = 0.$$

For computational purposes, we can let

$$Y_1 = 2\cos x + 5\sin x + 1.$$

The graphs of both are given in Figure 7-4 with the graph of Y_1 as a dotted line.

FIGURE 7-4

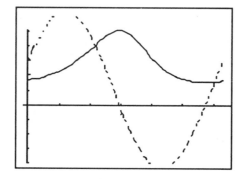

The derivative has zeros between 2 and 4 and between 4 and 6.

We use the modified ZERO program to approximate these; the screens are displayed in Figure 7-5.

FIGURE 7-5

```
?4
ZERO AT
          2.947865737
Y1=
          2.213E-9
Y2=
        5.097167541
```

```
?6
ZERO AT
          5.716899469
Y1=
          -3.049E-9
Y2=
        1.569499126
```

Between 2 and 4, the derivative is 0 at $x = 2.947865737$ and the local maximum value is $y = 5.097167541$.

Between 4 and 6, $y' = 0$ at $x = 5.716899469$ and the local minimum value is $y = 1.569499126$.

In this example, we only needed to examine the function over an interval of length 2π because the function is periodic with period 2π. The rest of the local extrema have the same y values and occur at numbers obtained by adding even multiples of π to the previously obtained values for x.

Maximum value: $y = 5.097167541$ at $x = 2.947865737 + 2n\pi$

Minimum value: $y = 1.569499126$ at $x = 5.716899469 + 2n\pi$. ‖

If your calculator has a SOLVE key or other process for finding zeros, it can be used instead of the ZERO program to find the zeros of the derivative. We could use the TRACE and ZOOM method to find the zeros but then we would have no better accuracy than with the Graphics Approximation Method and computing the derivative would have gained us nothing.

To find the absolute maximum and minimum values for a function on a closed interval, we need to compare the local maximum and minimum values and may need to compute one or both of the function values at endpoints.

Example 3: Determine the absolute maximum and minimum values of
$$y = x^3 - 3x^2 - 3x + 2 \quad \text{for } 0 \le x \le 4$$

SOLUTION: First of all we draw the graph with 0 and 4 as the minimum and maximum values of x in the RANGE setting. We can make a first estimate for y values and then revise it if necessary. We try [0,4,.5,-10,10,1,1] and it is sufficient. The graph is given in Figure 7-6 with Y_2 as the function.

FIGURE 7-6

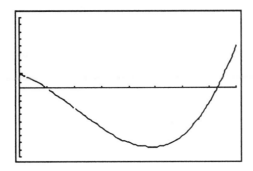

It appears that the maximum value occurs at the right endpoint and that the minimum will occur at the local minimum.

The derivative is

$$y' = 3x^2 - 6x - 3$$

The graphs of the functions are drawn together in Figure 7-7 with the graph of the derivative as a dotted line.

FIGURE 7-7

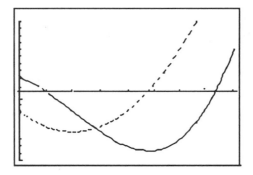

From the graph, we see that the derivative has just one zero between 0 and 4. The results obtained with the **ZERO** program are displayed in Figure 7-8.

FIGURE 7-8

```
?4
 ZERO AT
      2.414213663
Y₁=
          -3.83E-9
Y₂=
   -8.656865424D
```

The derivative has a zero at x = 2.414213562 and the minimum value of the function at that point is y = -8.856854249.

From Figure 7-7, we see that the maximum occurs at 4, the right endpoint, and we can use the TI-81 calculator to find this for us. Recall that to evaluate the function Y_2 at x = 4, you store 4 in X, display Y_2 and press ENTER. The number 6 is displayed as the value of Y_2 on the screen in Figure 7-9.

FIGURE 7-9

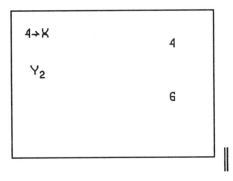

If it is not clear from the graph that an extrema occurs at an endpoint, we must compute the function values at both endpoints and compare them to the local extrema.

Determining maximum and minimum values has many applications in business, science, and engineering.

Example 4. Based on data, suppose you found that the profit y produced by a bagel shop each week is approximated by the function

$$y = 100[8x^2 - 2x^3 + x^{-3} - 1.25x] - 400 \quad \text{for } 1 \le x \le 4$$

where x is the price per dozen.

What would you advise the owner to set as a price in order to maximize the profit?

SOLUTION: Draw the graph as in Figure 7-10 with a RANGE setting of $[1,4,1,-400,2000,200,1]$. Notice that the function has one local maximum and that this is the maximum on the interval $[1,4]$.

FIGURE 7-10

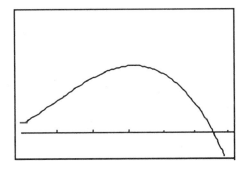

The derivative is

$$y' = 100[16x - 6x^2 - 3x^{-4} - 1.25]$$

Both graphs are given in Figure 7-11 with the graph of the derivative as a dotted line.

FIGURE 7-11

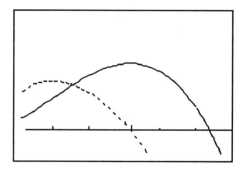

The derivative has only one zero and it is between 2 and 3. The results of applying the **ZERO** program are displayed in Figure 7-12.

FIGURE 7-12

```
?3
ZERO AT
         2.581607257
Y₁=
            6.306E-7
Y₂=
         1173.742441
```

The maximum profit is $1173.74 at a price of $2.58.

It is also valuable to see what the profit would be at prices of **$2.50**, **$2.55** and **$2.60** because these are more common "commercial" prices. We leave finding these numbers as an exercise. ‖

Example 5. Find the point on the graph of

$$\frac{x^2}{9} + \frac{y^2}{4} = 1$$

that is closest to the point (1,5).

SOLUTION: Since the point is above the x-axis, we only need be concerned with the upper portion of the curve. When we solve the equation for y we get

$$y = [4 - (4/9)x^2]^{1/2}$$

The graph is given in Figure 7-13.

FIGURE 7-13

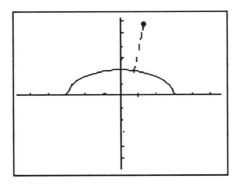

The distance from the curve to the point (1,5) is the minimum value of

$$Y_1 = [(x-1)^2 + ([4-(4/9)x^2]^{1/2} - 5)^2]^{1/2}$$

The derivative is quite complicated but the Graphics Aproximation method yields an approximate solution of 3.0665909 after only two applications of TRACE and ZOOM starting with a RANGE setting of [-3,3,1,1,5,1,1]. The graphs are given in Figure 7-14.

FIGURE 7-14

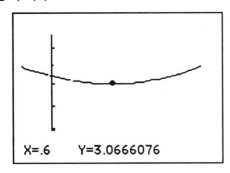

X=.6   Y=3.0666076

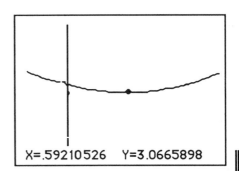

X=.59210526   Y=3.0665898

With a graphics calculator, many maximum and minimum problems in science and engineering can be solved once the quantity to be maximized or minimized is expressed as a function even if the functions are quite complicated and do not fall to the usual techniques of calculus. The skills involved in expressing such quantities are even more important now than they were before the power of the graphics calculator and that of the computer were available.

EXERCISES:

1. Use the Graphics Method to approximate the other local extrema in Example 1.

2. Use the Graphics Method to approximate the local extrema of
$$y = x^3 - 3x^2 - 6x + 7$$

3. Use the Graphics Method to approximate the local extrema of
$$y = 2\sin(3x) + 3\cos(4x)$$

4. Use the Graphics Method to approximate the local extrema of
$$y = \frac{x^3 + 5x - 7}{x^4 + 7}$$

In Exercises 5-14, use the Computational Method to approximate the local extrema of the functions. When possible, determine any absolute extrema.

5. $y = x^4 + 3x^3 - 18x^2 - 6x + 2$

6. $y = \sin(x\cos x)$ for $-\pi \le x \le \pi$

7. $y = \dfrac{x}{1.5 + \sin x}$ for $-2\pi \le x \le 2\pi$

8. $y = \dfrac{x^3 + 5x^2 - 3x + 2}{x^4 + 1}$

9. $y = \dfrac{\tan x}{x^2}$ for $0 \le x \le 1.5$

10. $y = \sin(x/x^2 + 1)$ for $0 \le x \le 20$

11. $y = 5\sin(3x) + 4\cos(2x)$ for $0 \le x \le 2\pi$

12. $y = 3\tan(x/4) + 2\cos x$ for $-\pi \le x \le \pi$

13. $y = 2x\sin(x^2) + .7\cos(x^3)$ for $-2 \le x \le 1$

14. $y = \dfrac{x^2\sin x}{x^4 + 1}$ for $-3 \le x \le 3$

15. Determine the profit levels for prices of \$2.50, \$2.55, and \$2.60 for Example 4. On the basis of this information, what do you think the price per dozen should be? Why?

16. Suppose that an object is oscillating with the height y at time x in seconds given by
$$y = 3\sin(2x) + 4\cos(3x+1) + 5$$
What is the highest position of the object and what is the lowest?
At what times?

17. Suppose you build an office and a storage building as in the diagram in Figure 7-15. You wish to run water lines to both buildings from one tap into the main line. Where should the tap be located in order to minimize the amount of water line needed?

FIGURE 7-15

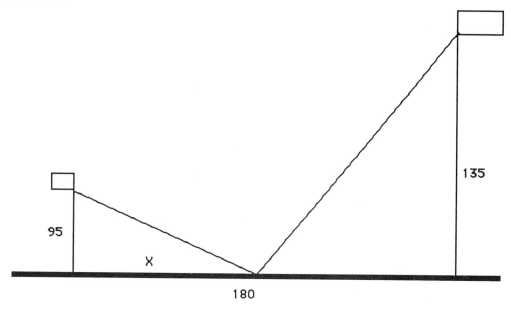

95

X

180

135

18. The intensity of light at a point is inversely proportional to the square of the distance of the point from the source. Suppose P and Q are two light sources 45 feet apart, M a point between them, and the source P is 5.7 times as bright as Q. If x is the distance of M from P, the intensity of the light at M is given by

$$I(x) = k\left[\frac{5.7}{x^2} + \frac{1}{(45-x)^2}\right]$$

for some constant k. Determine the point of minimum intensity and the intensity at that point.

19. Determine the minimum distance from the point (2.3) to the graph of $y = 3x^2 + 5x - 10$.

20. What point on the graph of

$$\frac{(x-10)^2}{9} + \frac{(y-12)^2}{16} = 1$$

is closest to the origin?

21. Suppose that the population of bacteria in a culture is given by

$$y = 10000(5x^4 + 7x^{-3})$$

At what time is the *rate of growth* the greatest?

22. Can you use your graphics calculator with the Graphics Method to find any local extrema for

$$y = 500x + \sin(1000x)$$

Can you do it with the Computational Method? Can you do it with standard calculus techniques? Which is the simplest? Explain your answers.

8

INCREASING AND DECREASING FUNCTIONS

When studying a particular function, we frequently wish to know the intervals where it is increasing and where it is decreasing. Two ways are available to us with the graphics calculator: one uses strictly graphics techniques and the other uses computational procedures.

GRAPHICS METHOD:

We can approximate the intervals where a function is increasing and those where it is decreasing by applying TRACE and ZOOM to the graph of the function,

Example 1. Determine where the function

$$y = \frac{x^2 - 5x - 8}{x^2 - 1}$$

is increasing and where it is decreasing.

SOLUTION: The graph is drawn twice in Figure 8-1; the first drawing uses the Standard RANGE and the second [-3,3,1,-20,20,5,1].

FIGURE 8-1

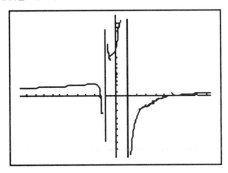 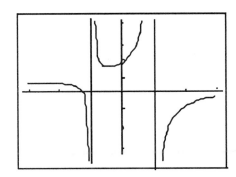

The behavior is not immediate from these graphs but the function does seem to be increasing from the left up to a point and then drops as the curve approaches the vertical asymptote.

If we apply TRACE and move the cursor to the left of the "hump" on the first branch of the curve , then moving the cursor to the right yields a value of y = 2.0504808 at x = -2.368421 with smaller values of y to the right and left. This is illustrated by the first graph in Figure 8-2. If this is close enough for our purposes, we can quit. If not, we can apply ZOOM and TRACE again as in the second drawing in Figure 8-2 to obtain x = -2.376316 and y=2.0505075. One more application of ZOOM and TRACE yields x = -2.378289 and y = 2.0505097 as in the third drawing in Figure 8-2.

FIGURE 8-2

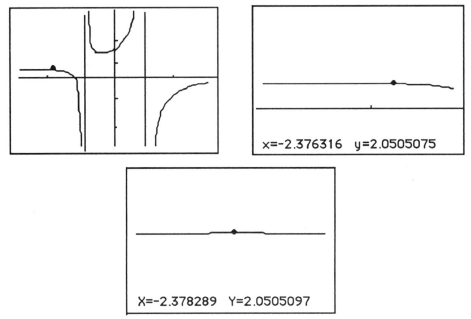

x=-2.376316 y=2.0505075

X=-2.378289 Y=2.0505097

The last two approximations are the same to two decimal places for x and five places for y; the last digits will vary with the choice of RANGE.

The function is decreasing as the graph approaches the vertical asymptote from the left and the cursor "jumps" from one branch to another at the asymptote x = -1. It is also decreasing to the right of the asymptote.

Looking at the original graph again in the first drawing in Figure 8-3, we see that the graph appears to have a low point. An application of TRACE gives a first estimate of x= -.4105263 and y = 6.9501599.

FIGURE 8-3

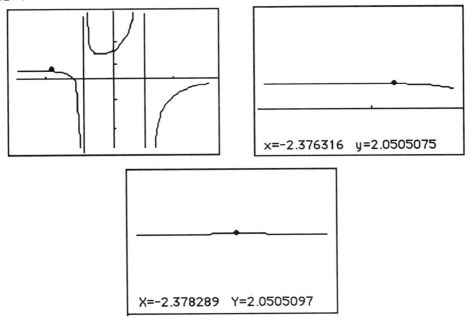

x=-2.376316 y=2.0505075

X=-2.378289 Y=2.0505097

Applying ZOOM and TRACE in the second drawing of Figure 8-3 gives a better estimate of x = -.4184211 and y = 6.9495127. A third application yields x = -.4203947 and y = 6.94949.

As before, we cannot expect any better accuracy.

In the original graph in Figure 8-1, we see that the graph is increasing to the left of the asymptote x = 1 and also is increasing to the right of it. It also appears to be increasing at all points as x moves toward the right. You can test this by several applications of TRACE.

From our work with rational functions, we know that the line y = 1 will be a horizontal asymptote.

From the above information, we know that the function is:

      increasing on (-∞,-2.375289]

      decreasing on [-2.375289,-1)

      decreasing on (-1,-.4203947]

      increasing on [-.4203947,1)

      increasing on (1,∞)

We include endpoints only when the function is defined at the point. ▌

      The graphics techniques serves our purpose very well in problems in which we need only a reasonable estimate. This occurs frequently in applications.

Example 2. Suppose that a concern finds that its profits y are approximated by

$$y = 10x - .001x^2 + 20x^{-3}$$

where x is the number of units produced.

For what levels of production is the profit increasing?

SOLUTION: In this case, the graphics techniques will suffice because the answer must be a whole number and we cannot even expect accuracy to the nearest integer. (A change of a very few units will not change the profits significantly.) The graph is given in Figure 8-4 with a RANGE setting of [100,10000,1000,0,30000,10000,1].

FIGURE 8-4

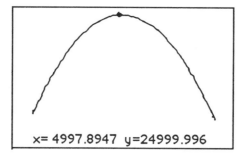

x= 4997.8947 y=24999.996

Using TRACE we see that the profits seem to increase as production increases from 0 to about 5000 units and to decrease after that. ▌

COMPUTATIONAL METHOD:

From calculus we know that a function $y = f(x)$ is increasing when its derivative is positive and decreasing when its derivative is negative. It may be easier to tell when the derivative is positive or negative than to tell when the function is increasing or decreasing. A method for finding zeros of functions can be very useful in such problems.

Example 3. Determine when the function

$$y = x^5 - 2x^4 + 3x^3 + 6x^2 - 7x + 3$$

is increasing and when it is decreasing.

SOLUTION: First of all we draw the graph in Figure 8-5 with RANGE [-5,5,1,-10,20,5,1]. List the function as Y_2 to simplify our later work.

FIGURE 8-5

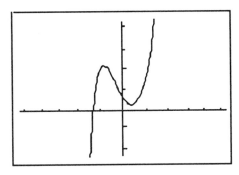

The derivative of the function is
$$y' = 5x^4 - 8x^3 + 9x^2 + 12x - 7$$
The graphs of both are given in Figure 8-6 with the derivative entered as Y_1 and drawn as a dotted line. The RANGE setting is
[-3,3,1,-10,20,1,1]

FIGURE 8-6

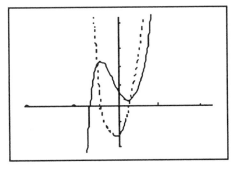

The derivative appears to have zeros between -2 and 0 and between 0 and 2. We can use the ZERO program to approximate these; the results are displayed in Figure 8-7.

FIGURE 8-7

```
?-2
ENTER B
?0
ZERO AT
        -.9347965545
Y₁=
            -1.88E-9
```

```
?0
ENTER B
?2
ZERO AT
        .467590624
Y₁=
            7.298E-9
```

The function is:

increasing on $(-\infty, -.9347965545]$,

decreasing on $[-.9347965545, .467590624]$, and

increasing on $[.467590624, \infty)$. $\|$

The technique in the last example uses the graph of the function to determine whether the function is increasing or decreasing but we could reach our conclusions by studying the graph to see where the derivative is positive and where it is negative.

EXERCISES

1. Use the Graphics Method to determine where the function
$$y = x^3 - 5x^2 - x + 3$$
is increasing and where it is decreasing. Obtain results correct to three decimal places.

2. Use the Graphics Method to determine where the function
$$y = x^4 + x^3 - 5x^2 - 2x + 4$$
is increasing and where it is decreasing. Obtain results correct to two decimal places.

3. Use the Graphics Method to determine where the function
$$y = \frac{x^2 + 4x + 7}{x^2 - 4}$$
is increasing and where it is decreasing. Obtain results correct to two decimal places.

4. Use the Graphics Method to determine where
$$y = x\sin(4x) \quad \text{for } 0 \leq x \leq 4$$
is increasing and where it is decreasing. Obtain results correct to two decimal places.

5. Suppose the graph of the derivative of a function is given in

Figure 8-8. What can you say about the function?

FIGURE 8-8

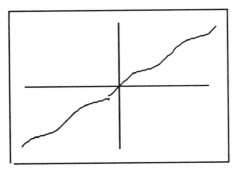

6. Suppose the graph of the derivative of a function is given in Figure 8-9. What can you say about the function?

FIGURE 8-9

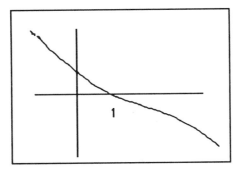

7. Use the Computational Method to determine where
$$y = x^3 - 4x^2 - 10x + 5$$
is increasing and where it is decreasing.

8. Use the Computational Method to determine where
$$y = \sin\left[\frac{x - 1}{x^2 - 4}\right]$$
is increasing and where it is decreasing.

9. Use the Computational Method to determine where
$$y = x + \sin(3x)$$
is increasing and where it is decreasing on $[0, 2\pi]$.

10. Use the Computational Method to determine where

$$y = 2x + \cos(7x)$$

is increasing and where it is decreasing on $[0, \pi]$.

11. Determine where

$$y = x^2 + 10\sin x$$

is increasing and where it is decreasing on $[0, 2\pi]$.

12. Determine where

$$y = \frac{x - 2}{x^2 + 1}$$

is increasing and where it is decreasing.

13. Determine where

$$y = \frac{x^2 - 7}{x^2 - 19}$$

is increasing and where it is decreasing.

14. Determine where

$$y = x^3 \tan x$$

is increasing and where it is decreasing.

15. Suppose that an object is traveling in a linear path with its distance from the starting point given by

$$s(t) = t^5 - 10t^3 + 17t + 3 \quad \text{for } 0 \leq t.$$

For what values of t is the distance from the starting point increasing?

16. Suppose an object is oscillating up and down with its height given by

$$h(t) = 7\sin(3t) + 5\cos(4t) + 100$$

When is it moving upward and when is it moving downward?

17. Can a non-constant even function be increasing? Explain your answer.

18. Can an increasing function have a horizontal asymptote? Justify your answer.

9

CONCAVITY

In calculus we learn that the graph of a function is concave upward on intervals where the second derivative is positive, concave downward on intervals where it is negative, and that inflection points are points on the graph where the concavity changes. If the second derivative exists at an inflection point, it must be zero. The graphics calculator enables us to get some notion of the concavity just by looking at the graph of the function.

Example 1. Determine the concavity of

$$y = x\sin x \qquad \text{for} \qquad -4\pi \leq x \leq 4\pi$$

SOLUTION: First draw the graph using RANGE setting [-12.6,12.6,4,-13,10,4,1].

To make our work simpler later on, we let

$$Y_2 = x\sin x$$

The graph is given in Figure 9-1.

FIGURE 9-1

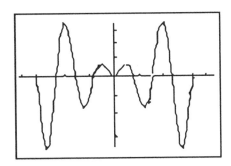

From the graph, it appears that the graph is symmetric with respect to the y-axis. Replacing x by -x and applying some algebra and trigonometry shows that the new function is the same and the symmetry holds. Thus to solve our problem, we only need study the behavior for values of x ≥ 0. We can reset the RANGE to [0,12.6,2,-13,10,4,1] and redraw the graph as in Figure 9-2.

FIGURE 9-2

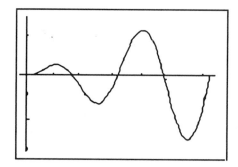

To the eye, the concavity seems to change at the zeros of the function and these are the points where $\sin x = 0$. We can use the **ZERO** program and the second derivative to see where the points of inflection really are.

First of all, $\qquad y' = x\cos x + \sin x$

and $\qquad y'' = -x\sin x + 2\cos x$

By drawing the graphs of y and y'' on the same system in Figure 9-3 (Let $y'' = Y_1$), we see that the zeros of y'' are not the same as the zeros of y and so our first conjecture about the inflection points does not hold.

FIGURE 9-3

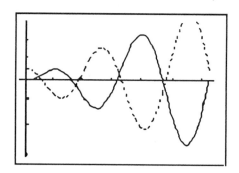

We can, however, use the **ZERO** program to determine the zeros of y''. In this case, the original graph is accurate enough to enable us to determine the concavity once we have this information. In fact, the concavity changes at each zero of y''. With a little editing, we can use the **ZERO** program to determine the inflection points. We just need to add $\text{Disp "}Y_2\text{="}$ and $\text{Disp } Y_2$. (Notice that finding the zeros of y'' algebraically would be a difficult problem.) From the graph of y'' (i.e. Y_1) we can see that t y'' has zeros between 0 and 2, 2 and 4, 6 and 8, 8 and 10. The edited **ZERO** program determines the zeros and the corresponding y values. The results are displayed in Figure 9-4.

FIGURE 9-4

```
?2
ZERO AT
        1.076873987
Y₁=
            -6.498E-10
Y₂=
    .9481661382
```

```
?4
ZERO AT
        3.643597167
Y₁=
        1.32E-10
Y₂=
    -1.753239569
```

```
?8
ZERO AT
        6.578333733
Y₁=
        -3.062E-9
Y₂=
    1.91351796
```

```
?10
ZERO AT
        9.629560343
Y₁=
        1.726E-9
Y₂=
    -1.958210524
```

The inflection points are:

$$(1.076873987, .9481661382)$$
$$(3.643597167, -1.753239569)$$
$$(6.578333733, 1.91351796)$$
$$(9.6295560343, -1.958210524)$$

From the drawing and this information, we now know that the graph is

concave down on $[0, 1.076873987]$,

concave up on $[1.076873987, 3.643597166]$,

concave down on $[3.643597166, 6.578333733]$,

concave up on $[6.578333733, 9.6295560343]$ and

concave down on $[9.6295560343, 12.6]$

Since the graph is symmetric to the y-axis, the same concavity holds on the corresponding negative intervals, e.g. the graph is concave down on $[-12.6, -9.6295560343]$. ‖

Example 2.   Determine the concavity of

$$y = \cos 2x + \sin x \quad \text{for } 0 \leq x \leq 2\pi$$

SOLUTION: If we draw the graph in Figure 9-5 letting $y = Y_2$ and use the RANGE setting [0,6.3,1,-4,4,1,1] it appears that the concavity changes four times.

FIGURE 9-5

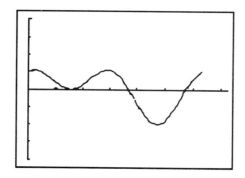

If we let

$$Y_1 = y'' = -4\cos 2x - \sin x$$

and draw this on the same coordinate system in Figure 9-6, we see that the second derivative has four zeros: one between 0 and 1, one between 2 and 3, one between 3 and 4 and one between 5 and 6.

FIGURE 9-6

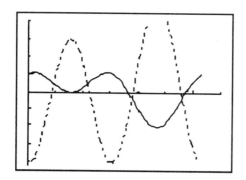

The ZERO program can be used to find these and the corresponding values of y; the results are displayed in Figure 9-7.

FIGURE 9-7

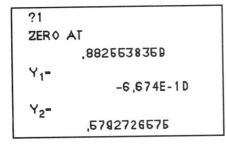

From the drawing and this information, we see that the graph is
concave down on $[0,.8825538359\}$
concave upward on,$[.8825538359,2.259038818]$
concave downward on $[\ \ 2.259038818,3.845712893]$
concave upward on $[3.3845712893,\ \ 5.579065067]$
concave downward on $[5.579065067,2\pi]$. ‖

EXERCISES

1. Determine the concavity of
$$y = x^3 + 5x^2 - 7x + 3$$

2. Determine the concavity of
$$y = 2x^5 - 7x^4 + 3x^3 + 5x^2 - 7x + 3$$

3. Determine the concavity of
$$y = \frac{10}{x^2 + 7}$$

4. Determine the concavity of
$$y = \frac{4x}{x^2 + 2}$$

5. Determine the concavity of
$$y = x^2\sin(2x) \qquad \text{for } 0 \leq x \leq 2\pi$$

6. Determine the concavity of

$$y = \sin(\cos x) \qquad \text{for } 0 \le x \le 2\pi$$

7. Determine the concavity of

$$y = \tan\left(\frac{1}{x^2 + 1}\right) \qquad \text{for } -1 \le x \le 1$$

8. Determine the concavity of

$$y = \cos(\sin(2x)) \qquad \text{for } 0 \le x \le 2\pi$$

9. Determine the concavity of

$$y = (2 + \sin x)^2 \qquad \text{for } 0 \le x \le 2\pi$$

10. Determine the concavity of

$$y = \frac{1 + \cos x}{2 + \sin x} \qquad \text{for } 0 \le x \le 2\pi$$

11. Determine the concavity of

$$y = \frac{\tan x}{x^2 + 1} \qquad \text{for } 0 \le x \le \pi$$

12. Determine the concavity of

$$y = x^4 + 5x^2 - x + 3$$

Explain and generalize your result.

13. Determine the concavity of a function if the second derivative is given by

$$y'' = x^2 - 6x - 10$$

Why is this information sufficient?

14. Determine the concavity of

$$y = \sin(3x) \quad \text{and} \quad y = \sin(3x) + 7x + 3$$

Is there anything special about your answers? Explain it and generalize your conclusion.

EXPLORATION AND EXTENSION

15. Determine the concavity of
$$y = p(x)$$
for some polynomial $p(x)$ of your choice but with odd degree of at least 3. Does the concavity change? Make a general conclusion and justify it.

16. Determine the concavity of
$$y = 2 - \cos x \text{ and } y = (2 - \cos x)^2 \text{ on } [0,1]$$
Try several examples to reach a general conclusion about the concavity of certain functions and their squares. State and justify your conclusion.

1 0

NEWTON'S METHOD

Some functions have the property that for values of x close to a zero of the function, the tangent line to the curve at that point intersects the x-axis near the zero. Each time the process is repeated, the approximation gets closer to the zero. This process for approximating zeros is called Newton's Method. The method fails occasionally but is valid for a large class of functions. We will not attempt to carefully describe those functions for which it is valid but refer you to any calculus book. We provide a graphic description of the method and a program for its use.

The drawing in Figure 10-1 illustrates the first three steps of Newton's Method.

FIGURE 10-1

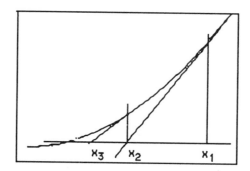

The initial value of x is denoted by x_1 and is chosen as a reasonable first approximation for the zero. Then x_2 is the point where the line tangent to the curve at $(x_1, f(x_1))$ meets the x-axis. Now draw the tangent line at $(x_2, f(x_2))$ and x_3 is the point where this line meets the x-axis. We continue the process until we have a sufficiently close approximation.

The formula for obtaining x_{n+1} from x_n can be derived using some

algebra and the fact that the derivative is the slope of the tangent line.

$$x_{n+1} = x_n - \frac{f(x_n)}{f'(x_n)}$$

Example 1. Find the second approximation for a zero of $f(x)$ using Newton's Method if

$$f(x) = 4x + \cos(\pi x) - 2.5$$

and the first approximation is $x_1 = 1$.

SOLUTION: The derivative is

$$f'(x) = 4 - \pi\sin(\pi x)$$

Computing yields

$$f(1) = .5$$
$$f'(1) = 4$$

Using the formula, we have

$$x_2 = x_1 - \frac{f(x_1)}{f'(x_1)}$$

$$x_2 = 1 - \frac{.5}{4} = .875$$

The geometry associated with the computation is illustrated in Figure 10-2.

FIGURE 10-2

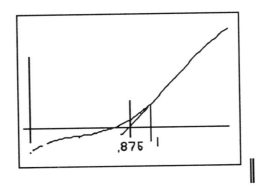

In order to use the method to approximate zeros, you need to compute successive approximations until $\left| x_n - x_{n+1} \right| < E$ where E is the desired

degree of accuracy. In the following program you need to enter the degree of accuracy required and make the initial choice of a value of x near a zero. The program is called **NEWTON**. Before using the program, enter the function as Y_1 and its derivative as Y_2. If you want accuracy to a certain number n of decimal places, it usually is simplest to let $E = 10^{-(n+1)}$; e.g. if you want four place accuracy, let $E = .00001$.

```
Prgm :NEWTON              :A→X
:Disp "ENTER E"           :Goto 1
:Input E                  :Lbl 2
:Disp "ENTER X"           :Disp "SOLUTION
:Input X                  IS"
:Lbl 1                    :Disp A
:X-(Y₁/Y₂)→A              :Disp "F(A) ="
:If abs (X-A)<E           :Disp Y₁
:Goto 2                   :End
```

Example 1.   (Revisited)

SOLUTION: We will compute an approximation correct to three decimal places.

First of all, let $Y_1 = 4X + \cos(\pi X) - 2.5$

$$Y_2 = 4 - \pi\sin(\pi X)$$

We will enter E as .0001.

The results of applying **NEWTON** are displayed in Figure 10-3.

FIGURE 10-3

```
?.0001
ENTER X
?1
SOLUTION IS
        .8464842055
F(A)=
        7.454578E-5
```

The solution is approximately .846. We rounded because we chose E = .0001 and we should not expect that the rest of the digits are accurate. ‖

Example 2. Use Newton's Method to approximate a solution of
$$x^3 - 7x^2 + 5x - 6 = 0$$

to within five decimal places.

SOLUTION: Let

$$Y_1 = X^3 - 7X^2 + 5X - 6$$

Then the derivative is
$$Y_2 = 3X^2 - 14X + 5$$

From the graph of Y_1 given in Figure 10-4, we see that 6 is a

reasonable initial estimate.

FIGURE 10-4

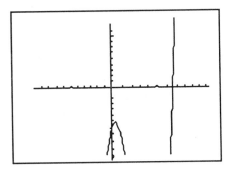

We enter Y_2 and apply **NEWTON.** We choose E = .000001 because we

want five place accuracy. The results are displayed in Figure 10-5

FIGURE 10-5

```
?000001
ENTER X
?6
SOLUTION IS
          6.362350043
F(A)=
          8.43129E-6
```

The solution is approximately 6.36235. ‖

Newton's Method can be combined with numerical differentiation by

letting Y_2 be the numerical approximation for $f'(x_n)$. To do so let

$$Y_2 = \text{NDeriv}(Y_1, \Delta x)$$

for some choice of Δx.

Example 3. Use Newton's Method to approximate the smallest positive zero
of

$$f(x) = [\sin(\cos(x^3 + 3x - 2))]^2$$

Give a solution correct to two decimal places.

SOLUTION: First let

$$Y_1 = [\sin(\cos(x^3 + 3x - 2))]^2$$

The graph is given in Figure 10-6 with RANGE [-1,1,1,-.1,1.1,1,1]

FIGURE 10-6

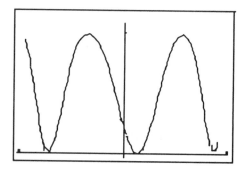

There appears to be a positive zero near 0.

For this example, let Δx = .01, Y_2 = NDeriv(Y_1,.01), and apply

NEWTON with E = .001. The results are displayed in Figure 10-7.

FIGURE 10-7

```
?.001
ENTER X
?0
SOLUTION IS
        .1411445912
F(A)=
        3.555546921E-5
```

The solution is approximately .14. ‖

It is natural to wonder why we introduced another method for finding
zeros of functions when we already had the **ZERO** program and especially
why we would introduce one that does not always work. One answer is
illustrated by Figure 10-6; a function may have zeros at points where it does

not cross the x-axis and the Bisection Method is not applicable.   Speed provides a second reason; Newton's Method usually gives the approximation in fewer steps and, therefore, in less time.   The third reason is tradition; the method was extremely valuable before the development of machines to do the calculations because it took fewer steps than bisection.

EXERCISES

1.   Use Newton's Method to compute the second approximation for a zero of
$$y = x^3 - 4x^2 + 7x - 2$$
if the first approximation is $x_1 = 1$.

2.   Use Newton's Method to compute the third approximation for a zero of
$$y = x^4 + x^3 + 2x - 1$$
if the initial value is $x_1 = -1$

3.   Use **NEWTON** to approximate a zero of
$$y = \sin(x^2 + x + 1)$$
to two decimal places if the initial value is $x_1 = 1$.

4.   Use **NEWTON** to approximate a zero of
$$y = \cos(x^2)$$
to two decimal places if the initial value is $x_1 = 1.5$.

5.   Use **NEWTON** to approximate the smallest positive solution of the equation
$$\sin(2x) + \cos(x+3) = 0$$
to within three decimal places.

6.   Use **NEWTON** to approximate the smallest positive solution of the equation
$$\tan(3x) = \cos(6x)$$
to within three decimal places.

7.   Use **NEWTON** to find each of the zeros of
$$y = \sin x + \cos x + .7$$
for $-2\pi \leq x \leq 2\pi$ to within five decimal places.

8.   Use **NEWTON** to approximate each of the zeros of
$$y = x + x^2\sin x - 1$$
for $0 \leq x \leq 2\pi$ to within five decimal places.

9.   Use **NEWTON** to solve the system of equations
$$y = x^2 - 3x$$
$$y = \sin x$$
to within four decimal places.

10.   Use **NEWTON** to solve the system of equations
$$y = x^3 - 4x^2 + 3x - 2$$
$$y = \cos x$$
to within four decimal places.

EXTENSION

11.   Use both **NEWTON** and **ZERO** to approximate any zeros of
$$y = x^3 + 5x - 11$$
Compare your results and the amount of calculator time each takes. Remember to expect the same degree of accuracy from **NEWTON** as is required in **ZERO**. (Hint: Look at the program for the value of E.)

12.   Try to use **NEWTON** to find a zero of
$$y = x^{.2}$$
Use geometry to explain why Newton's Method is invalid in this case.

1 1

NUMERICAL INTEGRATION: INTRODUCTION

Many integrals do not yield to the Fundamental Theorem of Calculus because the integrands do not have elementary antiderivatives or because finding such antiderivatives is too time consuming. Numerical techniques can be used to find approximations for such definite integrals and these methods expand our problem solving prowress.

PREREQUISITE SKILLS:

1. Knowledge of basic integration.
2. Elementary programming skills with your calculator.

The Trapezoidal Rule is based on approximating the integral by using trapezoids and is illustrated in Figure 11-1.

FIGURE 11-1

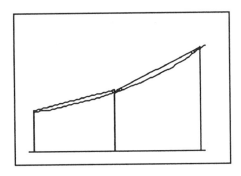

The formula is given by

$$\int_a^b f(x)dx \approx \frac{b-a}{2n} [f(x_0)+2f(x_1)+2f(x_2)+\cdots+2f(x_{n-1})+f(x_n)]$$

for $x_0 = a$, $x_1 = a+h$, $x_2 = a+2h$, $\cdots$, $x_j = a+jh$, $\cdots$ where $h = \dfrac{b-a}{n}$

A program for the Trapezoidal Rule is below. We suggest that you store it in your calculator.

To use the program, first enter the integrand as Y_1.

TRAPEZOIDAL RULE:

```
Prgm  :TRAP                    :A→X
:Disp"ENTER A,B,               :1→M
AND N, NUMBER OF D             :Lbl 2
IVISIONS"                      :X+D→X
:Disp"ENTER A"                 :Y₁+K→K
:Input A                       :1+M→M
:Disp "ENTER B"                :If M<N
:Input B                       :Goto 2
:Disp"ENTER N"                 :K*D→K
:Input N                       :Disp"APPROX
:(B-A)/N→D                     IS"
:A→X                           :Disp K
:Y₁/2→K                        :End
:B→X
:Y₁/2+K→K
```

Example 1. Use the program **TRAP** with 64 subdivisions to approximate

$$\int_0^1 \tan x^2 \, dx$$

SOLUTION: $Y_1 = (\tan x)^2$ and the values of A and B are 0 and 1.

The results are given by the screen displayed in Figure 11-2.

FIGURE 11-2

```
?0
ENTER B
?1
ENTER N
?64
APPROX IS
        .5576247707
```

The integral is approximated by .5576247707. ‖

Simpson's Rule is another common approximation formula. (n is even)

$$\int_a^b f(x)dx \approx \frac{b-a}{3n}\ [f(x_0)+4f(x_1)+2f(x_2)+4f(x_3)+\cdots+2f(x_{n-2})+4f(x_{n-1})+f(x_n)]$$

A program for Simpsons Rule is given below. This is the program provided by the manufacturer of the TI-81 and can be stored in your calculator. (We have changed the word "area" to "integral"; the integral might not be an area.)

Again, enter the integrand as Y_1 before executing the program.

SIMPSON'S RULE:

```
Prgm :SIMPSON          :0→S                :R→X
:All-Off               :(B-A)/2D→W         :Y₁→R
:Disp"LOWER LIM        :1→J                :W(L+4m+R)/3+S→S
IT"                    :Lbl 1              :IS>(J,D)
:Input A               :A+2(J-1)W→L        :Goto 1
:Disp"UPPER LIM        :A+2JW→R            :Disp "INTEGRAL ="
IT"                    :(L+R)/2→M          :Disp S
:Input B               :L→X                :End
:Disp"N DIVISIO        :Y₁→L
NS"                    :M→X
:Input D               :Y₁→M
```

Example 2. Use Simpson's Rule to approximate

$$\int_1^3 \sqrt{x^2+x+3}\ dx$$

SOLUTION: $Y_1 = \sqrt{x^2+x+3}$ with A = 1 and B = 3. We use 64 subdivisions. The results are displayed in Figure 11-3.

FIGURE 11-3

```
?1
UPPER LIMIT
?3
N DIVISIONS
INTEGRAL =
        6.035379342
```

The integral is approximated by 6.035379342. ▌

The Midpoint Rule approximates the integral by evaluating the function at the midpoints of the subdivisions, summing these values, and multiplying by the length of the subdivision. A geometric interpretation of for this rule is given in Figure 11-4.

FIGURE 11-4

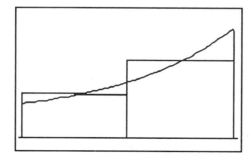

The formula for the Midpoint Rule is

$$\int_a^b f(x)\ dx \approx \frac{b-a}{n}\ [f(z_1) + f(z_2) + f(z_3) + \cdots + f(z_n)]$$

where $z_i = (x_{i-1} + x_i)/2$.

We do not provide a program here for the Midpoint Rule but ask you to do so as an exercise.

These approximation techniques can be used to solve problems in which we need to determine a definite integral but have no way of finding an antiderivative.

Example 3. Suppose that the rate of change of the sales of a business is known to be approximated by

$$y' = \frac{x + 50}{\sqrt{x + 1}}$$

where x is the time in years since 1950 and y is in thousands of dollars. Approximately what is the total change in sales in the 30 year period from 1960 to 1990?

SOLUTION: The total change is

$$\int_{10}^{40} \frac{x+50}{\sqrt{x+1}} \, dx$$

but we have no simple way to find an antiderivative for the function. We choose n = 128 and obtain approximations with both Simpson's Rule and the Trapezoidal Rule. Both screens are displayed in Figure11-5.

Remember that we first enter

$$Y_1 = \frac{x+50}{\sqrt{x+1}}$$

FIGURE 11-5

SIMPSON

```
?1D
UPPER LIMIT
?4D
N DIVISIONS
?12B
INTEGRAL -
     382.4779381
```

TRAPEZOIDAL

```
?1D
ENTER B
?4D
ENTER N
?12B
APPROX IS
     382.4793628
```

Each method shows a total increase in sales of about $382,500 over the thirty year period.   In fact it is probably more reasonable to say that the total increase is somewhere between $382,000 and $383,000. ‖

Area problems can be solved by combining these approximation techniques with other methods.

Example 4. Determine the area of the region bounded by the graph of
$$y = \sin(x^2) - x^2 + x + 1$$
and the x-axis.

SOLUTION: First we need to determine the region.   The graph of the function is given in Figure 11-6.

FIGURE 11-6

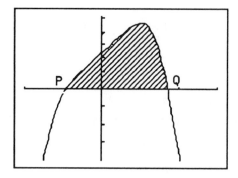

We see that the curve intersects the x-axis at two points.　The smaller x-coordinate is between -1.5 and 0 and the second is between 1 and 2. If we set

$$Y_1 = \sin(x^2) - x^2 + x + 1$$

the zeros of this function will be the x-coordinates of these points and the area will be the definite integral of the function between those points.　We denote the smaller by P and the larger by Q.

We first approximate P and Q and then approximate the area by applying the Trapezoidal Rule to

$$\int_P^Q [\sin(x^2) - x^2 + x + 1]\ dx$$

The **ZERO** program can be used to find the points where the curve meets the x-axis.　The results are given by the screens in Figure 11-7.

FIGURE 11-7

```
?0
ZERO AT
       -.9090853668
Y1=
       -7.617E-1D
M→P
       -.9090853668
```

```
?2
ZERO AT
        1.710292
Y1=
        6.061E-1D
M→Q
        1.710292
```

Notice that we have stored the values of the zeros for the limits of the integration and do not have to bother to copy them.　When the program says "ENTER A" just input P and similarly input Q for B.

Let $Y_1 = \sin(x^2) - x^2 + x + 1$ and use 128 subdivisions.　The results are given by the screens in Figure 11-8.

FIGURE 11-8

```
Prgm
ENTER A,B,AND N
NUMBER OF SUBDIVISI
ONS
ENTER A
?P
```

```
ENTER B
?Q
ENTER N
?12B
APPROX IS
        2.393238906
```

The area is approximately 2.393. ‖

When we use approximation techniques, it is natural to wonder just how good the approximations are. There are error bounds for both of the rules given above.

If we denote the error in applying the Trapezoidal Rule by E_T, then it can be shown that

$$\left| E_T \right| \leq \frac{(b-a)^3}{12n^2} M$$

where M is an upper bound for $\left| f'' \right|$ on the interval [a,b].

Similarly, if the error in applying Simpson's Rule is denoted by E_S, then

$$\left| E_S \right| \leq \frac{(b-a)^5}{180n^4} M$$

where M is any upper bound of $\left| f^{(4)} \right|$ on the interval [a,b].

The error bound for Simpson's Rule seems better for smaller values of n but using such a bound requires computing a fourth derivative whereas the bound for the Trapezoidal Rule only requires computing a second derivative.

Example 5. Use the Trapezoidal Rule to determine an approximation of

$$\int_0^1 \sin x^2 dx$$

with an error of no more than .001.

SOLUTION: The first derivative of
$$y = \sin x^2$$
is
$$y' = 2x\cos x^2$$
and the second is
$$y'' = -4x^2\sin x^2 + 2\cos x^2.$$

We know that $\left|\sin x^2\right| \leq 1$ and $\left|\cos x^2\right| \leq 1$ and since $x^2 \leq 1$, we can let $M = 6$ and solve the inequality

$$\frac{1^3}{12n^2}\, 6 < .001$$

$$n^2 > \frac{6}{.001} = 6000$$

$$n > 77.459$$

$$n \geq 78 \qquad \text{since } n \text{ is an integer.}$$

Now enter $Y_1 = \sin x^2$, $A = 0$, $B = 1$, and $n = 78$.

The results are displayed in Figure 11-9.

FIGURE 11-9

```
?0
ENTER B
?1
ENTER N
78
APPROX IS
     .3102831034
```

The integral is within .001 of .310. ‖

From the study of basic integration and the Fundamental Theorem of Calculus, we know that if the function $g(x)$ is continuous on the interval [a,b], then the function

$$f(x) = \int_a^x g(t)\ dt$$

has a derivative for each value of x, $a < x < b$, and that $f'(x) = g(x)$.

The approximation techniques developed earlier can be used to determine approximate values of such functions.

Example 6.   Suppose that the function $f(x)$ is given by

$$f(x) = \int_1^x \sqrt{t^3 + t + 1} \; dt$$

Use the Trapezoidal Rule to approximate $f(2.3)$; the graph of the integrand is given in Figure 11-10.

FIGURE 11-10

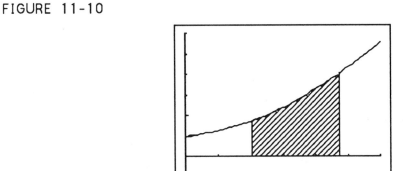

SOLUTION: Let $Y_1 = \sqrt{t^3 + t + 1}$, $A = 1$, $B = 2.3$, and $n = 128$. The results are displayed in Figure 11-11

FIGURE 11-11

```
?1
ENTER B
?2.3
ENTER N
?12B
APPROX I6
      3.643360343
```

In this case, $f(2.3) \approx 3.54335$. ‖

Some graphics calculators have built-in programs for approximating definite integrals.   The key may look something like $\boxed{\int f(x)dx}$ and can be used instead of the numerical approximation formulas given in this section.

EXERCISES

1. Approximate $\displaystyle\int_0^3 (x^2 + 2x + 3)\,dx$ with the Trapezoidal Rule and $n = 32$.

    Compute the integral using antidifferentiation and compare the results.

2. Approximate $\displaystyle\int_0^\pi \cos x\,dx$ with Simpson's Rule and $n = 32$.

    Compute the integral using antidifferentiation and compare the results.

3. Approximate $\displaystyle\int_0^2 \cos(x^3)\,dx$ with Simpson's Rule and $n = 64$.

4. Approximate $\displaystyle\int_0^3 \sqrt{x^2 + x + 1}\,dx$ with the Trapezoidal Rule and $n = 32$.

5. Approximate $\displaystyle\int_{-4}^4 \sin(x^3)\,dx$ with the Trapezoidal Rule.

6. Approximate $\displaystyle\int_1^4 \cos(x^2 + 1)\,dx$ with Simpson's Rule.

7. Use the Trapezoidal Rule to approximate $\displaystyle\int_1^2 \frac{1}{x}\,dx$ and $\displaystyle\int_1^8 \frac{1}{x}\,dx$ with $n = 64$.

    Compare your results.   Conjecture what $\displaystyle\int_1^{16} \frac{1}{x}\,dx$ might be.   Check it with

    the Trapezoidal Rule.

8. Use Simpson's Rule with $n = 128$ to approximate $\displaystyle\int_0^1 4\sqrt{1 - x^2}\,dx$. How does

    geometry explain your result?

9. Approximate the area of the region bounded by the graph of
$$y = \sqrt{17-5x^2-x^4}$$
and the x-axis.

10. Approximate the area of the region bounded by the graph of
$$y = \sin\left[\frac{1}{x^2+1}\right] - .3$$
and the x-axis.

11. Approximate the area bounded by the curve
$$y = \sqrt{.1x^4+x^2+1} - 4$$
and the x-axis.

12. Let $f(x) = \int_1^x \cos(t^2+4)\, dt$. For $1 \le x \le 3$, what value of n would assure that approximations of $f(x)$ using the Trapezoidal Rule are accurate to within .001?

13. Let $f(x) = \int_0^x \sin(t^3+2t^2+5)\, dt$. Approximate $f(.2)$, $f(.35)$ and $f(.9)$ within .01.

14. Let $f(x) = \int_0^x \sin(t^2)\, dt$. Estimate $[f(2.3) - f(2.2)]/.1$. Compare this with $y=\sin(t^2)$ evaluated at $t = 2.26$. Explain the similarity.(Hint: MVT)

15. Use the Trapezoidal Rule to approximate $\int_0^1 \cos x^3\, dx$ within .001. (Hint: Use the graph of the second derivative to find an upper bound.)

16. Use Simpson's Rule to approximate $\int_1^2 \cos(x^3)\, dx$ within .01.

(Hint: Use the graph of the fourth derivative to find an upper bound.)

17. Write a program for the Midpoint Rule.

18. Use your program for the Midpoint Rule to approximate

$$\int_0^2 x^2 \, dx$$

Compute the integral by antidifferentiation and compare the results.

19. Approximate $\int_0^{\sqrt{\pi}} \cos x^2 dx$ using the Midpoint Rule.

20. Suppose the graph of a function $f(x)$ is increasing on the interval $[a,b]$. Which seems to provide the better approximation for the integral,

$$\int_a^b f(x) \, dx$$

the Trapezoidal Rule or the Midpoint Rule? Give a geometric explanation for your conclusion.

21. Use TRAP to compute approximations of

$$\int_1^4 (x^3 - 5x^2 - 3x + 12) \, dx$$

for n = 32, 64, 128, and 256. Then compute it exactly using calculus. How does the accuracy change as n changes?

22. Choose $n = 4$ and use Simpson's Rule to approximate

$$\int_2^5 (x^3 - 9x^2 + 8x - 4) \, dx$$

Then compute the integral using antidifferentiation. How do the results compare? Use the discussion of Simpson's Rule in your calculus book to explain your results.

1 2

INVERSE FUNCTIONS

Two functions $f(x)$ and $g(x)$ are said to be *inverses* of each other if
$$f(g(x)) = x \text{ and } g(f(x)) = x$$
for all values of x where the expressions have meaning. The inverse of a function f is often denoted by f^{-1}.

Remember that $f^{-1}(x)$ does NOT mean $\dfrac{1}{f(x)}$.

Example 1. The functions

$$y = x^3 \text{ and } y = \sqrt[3]{x}$$

are examples of functions that are inverses of each other. Their graphs are given in Figure 12-1. Both domains and ranges are the set of all real numbers.

FIGURE 12-1

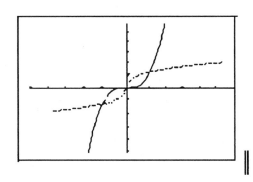

Example 2. The functions
$$y = x^2 \qquad \text{and} \qquad y = \sqrt{x}$$
are not inverses of each other since, although
$$\left[\sqrt{x}\right]^2 = x, \text{ in general, } \sqrt{x^2} \neq x.$$

In particular
$$\sqrt{(-5)^2} = 5 \neq -5 \;\|$$

If we modify the last example by restricting the values of x to positive numbers, then $y = \sqrt{x}$ is the inverse of $y = x^2$ for positive values of x. In such a case we say that $x = \sqrt{y}$ is a *partial inverse* for $y = x^2$.

The inverse trigonometric functions arcsine, arccosine, and arctangent are examples of partial inverses. For example, if $0 \leq y \leq \pi$ and $-1 \leq x \leq 1$, then y = arccos x if and only if cos y = x

The graph of the inverse of a function is the graph of the function reflected about the line y = x as illustrated in Figure 12-2.

FIGURE 12-2

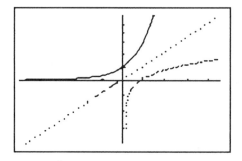

Since f sends x to y, f^{-1} should send y back to x and reflecting the graph about the line y = x does exactly that.

In calculus we learn that a continuous function that is strictly increasing or strictly decreasing has an inverse. Sometimes the graph of a function is enough to tell us if it is strictly increasing or strictly decreasing. Both of the functions whose graphs are given in Figure 12-3 have inverses.

FIGURE 12-3

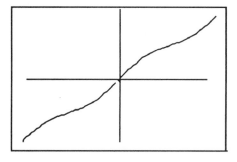 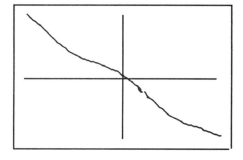

A more careful way of deciding whether or not a function has an inverse is to examine its derivative.

IN GENERAL:

1. If y' is positive, then y is increasing and has an inverse.

2. If y' is negative, then y is decreasing and has an inverse.

In fact, y' may have finitely many zeros and the conclusions will still hold. Thus if the graph of the derivative y' does not cross the x-axis and only touches it finitely many times, y has an inverse.

Example 3. Does the function
$$y = x^3 - 6x^2 + 16x - 15$$
have an inverse?

SOLUTION: The derivative is
$$y' = 3x^2 - 12x + 16$$

We could use algebra to determine if the derivative changes sign but it is simpler to draw the graph with RANGE [-5,5,1,0,50,10,1].

FIGURE 12-4

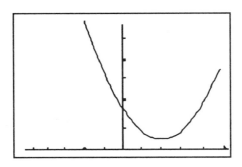

Since all points are above the x-axis, y' is positive for all values of x and y has an inverse. ‖

Finding an inverse for a function is frequently a hard problem but we can use the graphics calculator to determine good approximations. Remember that

if $y = f(x)$,   then $f^{-1}(y) = x$.

To find $f^{-1}(b)$ for $y = f(x)$:

GRAPHICS METHOD:

1. Draw the graph of $y = f(x)$.
2. Use TRACE to find an approximation of the the point where $y = b$.
3. The corresponding x value is an approximation of $f^{-1}(b)$.

Example 4. Determine an approximation of
$$f^{-1}(5)$$
if
$$f(x) = x^3 + x - 33.$$

SOLUTION: First of all, the derivative
$$f'(x) = 3x^2 + 1$$
is positive for all values of x and the function has an inverse.
If we draw the graph as in Figure 12-5 and use TRACE, we find that $x = 3.2631579$ when $y = 5.009914$. This approximation is very good and suffices for our purposes. We suggest that you see if you can do any better by applying TRACE and ZOOM.

FIGURE 12-5

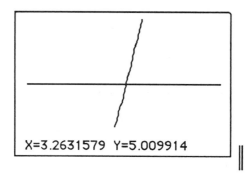

X=3.2631579 Y=5.009914

COMPUTATIONAL METHOD:

Solve the equation
$$f(x) = b$$
by solving the equation
$$f(x) - b = 0$$
To do so, let
$$Y_1 = f(x) - b$$
and apply the ZERO program.

Example 5. If g is the inverse of the function f(x) given by
$$y = f(x) = x^3 + 2x - 3$$
find an approximation of g(10).

SOLUTION: The inverse exists because
$$y' = 3x^2 + 2 > 0$$

Let $Y_1 = x^3 + 2x - 3 - 10$,
           $Y_1 = x^3 + 2x - 13$

and draw the graph with RANGE [-1,4,1,-20,20,10,1]
in Figure 12-6.

FIGURE 12-6

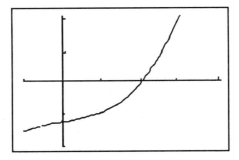

We see that there is a zero between 1 and 3. We use **ZERO** to find it and
the results are displayed in Figure 12-7.

FIGURE 12-7

```
?1
ENTER B
?3
ZERO AT
        2.069343933
Y₁ˉ
           -2.8E-1 D
```

The value of the inverse at 10 is approximately
$$g(10) \approx 2.069343933 \; \|$$

We can expect better accuracy from the computational method but frequently the results obtained using TRACE are sufficient.

Specific values of the derivative of the inverse of a function can frequently be found by using the following important result.

The Inverse Function Theorem:

If f(x) has an inverse g(y), f(a) = b, and f'(a) exists and is not zero, then

$$g'(b) = \frac{1}{f'(a)}$$

We can use this theorem and the graphics calculator to determine approximations for the values of the derivative of the inverse of a function without knowing the inverse or the derivative of the inverse.

Example 6. Let

$$f(x) = x^3 + x + 1$$

Find an approximation for g'(4) where g is the inverse of f.

SOLUTION: First of all:

$$f'(x) = 3x^2 + 1 > 0$$

and the inverse of f exists.

The first graph in Figure 12-8 is drawn with a RANGE of [-5,5,1,-10,10,5,1] and the second is the result of applying TRACE and ZOOM twice.

FIGURE 12-8

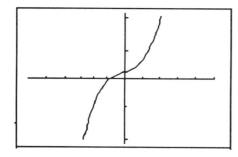 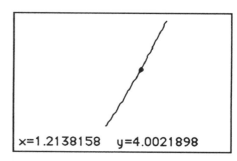

x=1.2138158   y=4.0021898

This method gives us x = 1.2138158 when y = 4.0021898; we store this value of x in X.

Now list the derivative as Y_2.

$$Y_2 = 3x^2 + 1.$$

By the Inverse Function Theorem, $g'(4)$ will be approximately $\dfrac{1}{Y_2}$

evaluated at the stored value of X. The results are displayed in

Figure 1 2 - 9

FIGURE 12-9

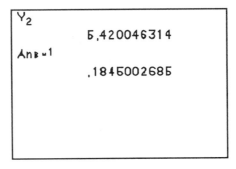

Therefore $g'(4) \approx .1845.$ ‖

We can also use ZERO program to find the desired value of x.

Example 7. Determine an approximation of $(f^{-1})'(7)$ if
$$f(x) = 2x + \sin x$$

SOLUTION: First of all, f does have an inverse because
$$f'(x) = 2 + \cos x > 0$$

We set $Y_1 = 2x + \sin x - 7$

and $Y_2 = 2 + \cos x$

Both graphs are drawn with the Standard RANGE in Figure 1 2 - 1 0.

FIGURE 12-10

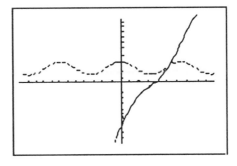

It appears that $Y_1 = f(x)$ has a zero between 3 and 5. The results of applying the **ZERO** program and completing the computation are displayed in Figure 12-11.

FIGURE 12-11

```
?5
ZERO AT
      3.809783432
Y₁=
        -1.93E-10
Y₂=
        1.215056115
1/Y₂
        .8230072558
```

Therefore $(f^{-1})'(7) \approx .82300$. ‖

There are many functions that do not have inverses. However, if we restrict the domain, we may obtain a partial inverse and in such a case we can still apply the inverse function theorem.

Example 8. Let

$$y = f(x) = 1 + x + 2\cos x$$

Show that f has a partial inverse g on an interval containing 6 and use the inverse function theorem to find $g'(6)$.

SOLUTION: First we draw the graph of

$$Y_1 = 1 + x + 2\cos x$$

in Figure 12-12 with the Standard RANGE setting.

FIGURE 12-12

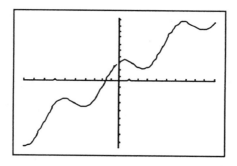

We can see that the function takes the value 6 and is increasing at that point. Thus the inverse exists on some interval containing 6.

Now we draw the graph of
$$Y_1 = 1 + x + 2\cos x - 6$$
i.e. $\qquad\qquad\qquad Y_1 = x + 2\cos x - 5$

in Figure 12-13.

FIGURE 12-13

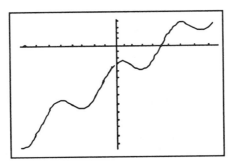

We can see that $y = f(x) - 6$ has a zero between 4 and 6.

Enter the derivative y' as
$$Y_2 = 1 - 2\sin x$$
and use the **ZERO** program to find the zero of Y_1. After computing the value of Y_2, we compute the value of $1/Y_2$ as before. The results are given in Figure 12-14.

FIGURE 12-14

```
?6
ZERO AT
        4.808357482
Y₁=
            3E-12
Y₂=
      2.980797113
1/Y₂
     .3343590228
```

For this partial inverse g, $g'(6) \approx .3343590228$. ‖

The inverse function theorem is a well known result but, except for special cases, using it requires numerical techniques such as those used here.

EXERCISES

1. Use the graph of

$$y = f(x) = x^3 + 4x - 7$$

   to decide whether or not f has an inverse.

2. Use the graph of

$$y = f(x) = \frac{x^3}{x^2 + 1}$$

   to decide whether or not f has an inverse.

3. Use the graph of

$$y = g(x) = 2x + \sin(3x)$$

   to decide whether or not g has an inverse.

4. Use the graph of

$$y = g(x) = x + 3\cos x$$

   to decide whether or not g has an inverse.

5. Use the graph of

$$y = f(x) = .01x^3 + .7x - 2$$

   and TRACE to approximate $f^{-1}(3)$.

6. Use the graph of

$$y = g(x) = 5x + \sin x$$

   and ZERO to approximate $g^{-1}(4)$.

7. Use the graph of

$$y = f(x) = \frac{x^2}{x^2 + 1}$$

   to determine two partial inverses for f. Use TRACE to approximate the two possible values for $f^{-1}(.3)$

8. Use the graph of
$$y = g(x) = (5x^2 - 3x - 4)^2$$
to determine two partial inverses for g. Use **ZERO** to find two possible values for $g^{-1}(14)$.

9. Use the inverse function theorem to approximate
$$g'(5) \text{ if } g = f^{-1} \quad \text{and} \quad y = f(x) = x^3 + 9x - 2$$

10. Use the inverse function theorem to approximate
$$g'(2) \text{ if } g = f^{-1} \quad \text{and} \quad y = f(x) = \frac{x^3}{x^2 + 1}$$

11. Use the inverse function theorem to approximate $(f^{-1})'(4)$ if
$$y = f(x) = x^3 + x + 1$$

12. Use the inverse function theorem to approximate possible values of $(p^{-1})'(6)$ if
$$y = p(x) = \frac{10}{x^2 + 1}$$

13. Determine two partial inverses for
$$y = f(x) = x^2 + 2x + 5$$
If g is one of these partial inverses, determine the possibilities for $g'(7)$.

14. If g is a partial inverse for
$$y = f(x) = x^3 - x + 2,$$
determine the possibilities for $g'(1.8)$

15. Determine four partial inverses for
$$y = f(x) = \frac{x^2 - 4}{x^2 - 9}$$
If g is one of these partial inverses, determine the possibilities for $g'(3)$.

16. Can an even function have an inverse? Can it have partial inverses? What about an odd function? Justify your answers.

1 3

LOGARITHMIC AND EXPONENTIAL FUNCTIONS

In algebra, logarithms are defined using exponents; in calculus, the natural logarithm is defined using integration.

DEFINITION: For any positive number x, the *natural logarithm* of x is denoted by ln(x) and defined by

$$\ln(x) = \int_{1}^{x} \frac{1}{t}\, dt$$

For x > 0, ln(x) is the area of the shaded region in the first drawing of Figure 13-1; for x < 0, it is the negative of the area of the shaded region in the second drawing.

FIGURE 13-1

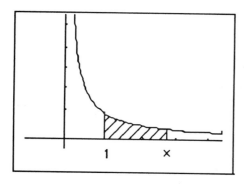
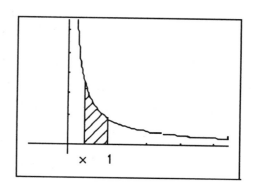

The number e is the unique positive number whose natural logarithm is 1 and therefore ln(e) = 1. This is illustrated in Figure 1 3 - 2.

FIGURE 13-2

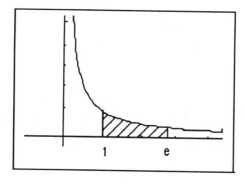

The *natural exponential function* is denoted by e^x and defined by

$$e^x = y \quad \text{if and only if} \quad \ln(y) = x$$

From the definition, we see that these functions are inverses of each other. The graphs are given in Figure 13-3. Notice that the y-axis is an asymptote for $\ln x$ and the x-axis is an asymptote for e^x.

FIGURE 13-3

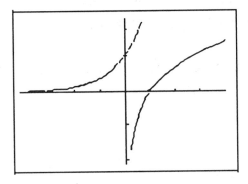

For any positive number b except 1, the logarithm to the base b of x is denoted by $\log_b(x)$ and defined by

$$\log_b(x) = \frac{\ln(x)}{\ln(b)}$$

The usual laws of logarithms and exponents hold; you are asked to illustrate and state some of them in the exercises.

The natural logarithmic and exponential functions are rather simple to use in calculus because

$$\frac{d}{dx}\ln(x) = \frac{1}{x} \quad \text{and} \quad \frac{d}{dx}e^x = e^x$$

These functions are so important that most calculators have special keys for them. For the TI-81, the natural logarithm key is

$$\boxed{\text{LN}}$$

and the natural exponential function e^x is obtained using

$$\boxed{\text{2nd}} \quad \boxed{\text{LN}} \; .$$

These functions can be combined with others to create functions that are rather complicated but the graphics calculator provides their graphs quite simply. Two such graphs are provided in Figure 13-4.

FIGURE 13-4

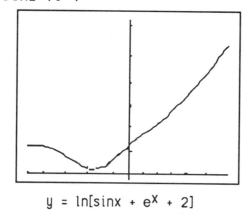 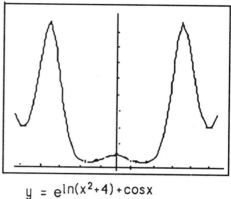

$$y = \ln[\sin x + e^x + 2]$$ $$y = e^{\ln(x^2+4)+\cos x}$$

Exponential functions provide quite good mathematical models of population   patterns.

Example 1. Suppose that it has been determined that the population for a particular city is given by

$$p = ce^{mt}$$

for some numbers c and m where p is the population and t is a time variable.

If you know that the population was 120,000 in 1970 and 123,500 in 1976, can you determine the function?   Estimate the population in 1985.

SOLUTION: If we use "stop watch" time and let t = 0 at the beginning of our observations, then

$$120000 = ce^0 = c$$

and the population is given by

$$p = 120000e^{mt}$$

for some number m.

We know that when t = 6, p = 123500.

Then $122500 = 120000e^{6m}$

$$e^{6m} = 12500/120000$$

$$6m = \ln(122500/120000) = \ln(1225/1220)$$

and $m = [\ln(1225/1200)]/6$

$$m = .0034365479$$

The population at time t after 1970 is given by

$$p = 120000e^{.0034365479t}$$

In particular, in 1985 it was approximately

$$p = 120000e^{(.0034365479)(15)}$$

$$p = 126348 \parallel$$

Logarithmic functions can be used to provide mathematical models of some commercial situations.

Example 2. Suppose studies in a particular industry have shown that the number of units y produced per hour is given by

$$y = 4.23\ln(x - 6)$$

where x is the hourly rate paid and the minimum paid to any worker is $8.00 per hour. If profit per unit is $.89, at what rate of pay does an increase in wages not result in an increase in profits?

SOLUTION: The graph of this function is given in Figure 13-5.

FIGURE 13-5

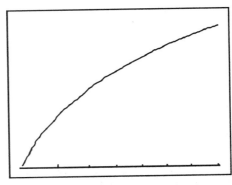

The rate of change of productivity is

$$\frac{d}{dx}[4.23(\ln(x - 6)] = \frac{4.23}{x - 6}$$

In order for profit to increase, you need

$$.89 < \frac{4.23}{x - 6}$$

$$x < 10.75$$

A rate of $10.75 or more does not result in an increase in profits. $\parallel$

Techniques developed earlier can be used with these functions as well.

Example 3. Compute $\int_{-1}^{3} e^{-x^2}\, dx$

SOLUTION: The graph of $y = e^{-x^2}$ is given in Figure 13-6.

FIGURE 13-6

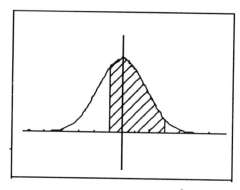

There is no simple antiderivative for e^{-x^2} but the integral can be approximated using one of the numerical techniques. We do so using Simpson's Rule with 128 subdivisions. The result is displayed on the screen in Figure 13-7.

FIGURE 13-7

```
?-1
UPPER LIMIT
?3
N DIVISIONS
?128
INTEGRAL =
       1.633031482
```

The integral is approximately 1.633. ‖

All of the techniques you learned with other functions can be applied to logarithmic and exponential functions. The ideas are the same, the particular functions are different.

EXERCISES

1. In each of the following use your calculator to compute the numbers and state the law of logarithms or exponents that it illustrates.

    (a) $\ln(5) + \ln(13)$, $\ln(65)$ (b) $\ln(135) - \ln(5)$, $\ln(27)$

    (c) $\ln(23^3)$, $3\ln 23$ (d) $e^3 e^4$, e^7

    (e) $\ln(e^{1.9})$ (f) $e^{\ln(5.7)}$

2. Draw the graphs of

    $$y = 2^x \qquad y = e^x \qquad y = 3^x \qquad y = 4^x$$

    on one coordinate system. How are the graphs related?

3. Draw the graphs of

    $$y = (.2)^x \qquad y = (.5)^x \qquad y = (.8)^x$$

    on one coordinate system. How are the graphs related?

4. Choose two numbers a and b with $1 < a < b$ and draw the graphs of

    $$y = a^x \qquad y = b^x$$

    on one coordinate system. How are the graphs related?

5. Choose two numbers a and b with $0 < a < b < 1$ and draw the graphs

    of $\qquad y = a^x \qquad$ and $\qquad y = b^x$

    on one coordinate system. How are the graphs related?

6. Draw the graphs of

    $$y = (.5)^x \qquad \text{and} \quad y = 2^{-x}$$

    On one coordinate system. How are the graphs related? Why?

7. Choose a number b between 4 and 10. Draw the graphs of

    $$y = b^{-x} \qquad \text{and} \quad y = \left(\frac{1}{b}\right)^x$$

    How are the graphs related? Give an algebraic explanation.

8. Draw the graphs of

    $$y = 3^{x+2} \qquad \text{and} \quad y = 9(3^x)$$

    How are the graphs related? Why?

9.   Draw the graphs of
$$y = \ln(5x) \quad \text{and} \quad y = \ln x + \ln 5$$
How are the graphs related?   Why?

10. Draw the graphs of
$$y = \ln(x^3) \quad \text{and} \quad y = 3\ln x$$
How are the graphs related?   Why?

11.   Draw the graphs of
$$y = \ln\left[\frac{x^2 + 7}{x + 3}\right] \quad \text{and} \quad y = \ln(x^2+7) - \ln(x+3)$$
How are the graphs related?   Why?

12.   Draw the graph of $y = \ln(x^2 - 4x - 5)$. Is there a break in the curve? Explain.

13.   Draw the graph of $\ln(2 + \sin x)$. Is there a break in the curve? Why?

14.   Draw the graph of $y = e^{\sin x}$. Is it periodic?   Why?

15.   Draw the graph of $y = \ln(x^2 + \cos x + 4)$. Is the function even? Why?

16.   When interest is compounded n times per year and r is the annual rate per year, the amount accrued after t years is given by
$$A = P\left(1 + \frac{r}{n}\right)^{nt}$$
where P is the original amount, the principal.

When interest is compounded continuously, the amount is given by
$$A = Pe^{rt}$$
Compute the amount accrued after 5 years if interest is compounded every three months with principal of \$15000 and an annual rate of .07. What would the amount be if the interest were compounded continuously?

17. We say that y *varies exponentially* with x if there are numbers c
     and m such that
     $$y = ce^{mx}.$$
     Determine c and m if $y = 1000$ when $x = 0$ and $y = 4000$
     when $x = 5$. Use this information to determine y when $x = 8$.

18. Using the language of Exercise 17, we can say that the amount varies
     exponentially with time when interest is compounded continuously. If
     you invested $2000 and received $7500 twenty years later with
     interest compounded continuously, what was the annual rate of
     interest?

19. The rate of decay of a radioactive substance is proportional to the
     amount of the substance. In calculus, it is shown that the amount A of
     such a substance remaining after time t is given by
     $$A = A_0 \left(\frac{1}{2} \right)^{\frac{t}{m}}$$
     where m is the half-life of the substance, the time it takes one half of
     the substance to decay. Let $Y = A$, $X = t$, and draw the graph of the
     amount of a substance if you start with 1000 grams, have a half life of
     20 years and time runs from 0 to 60 years. Use both TRACE and
     calculation to determine the amount after 37 years.

20. Use the Trapezoidal Rule with 128 subdivisions to approximate
     $$\int_2^4 \ln(x^2 + 3) \, dx$$

21. Use the Midpoint Rule with 64 subdivisions to approximate
     $$\int_1^3 \sin(e^{x^2}) \, dx$$

22. Use numerical differentiation with $\Delta x = .01$ to approximate $f'(1.2)$
     if $f(x) = \cos(x^2 + \ln x)$.

23.   Use numerical integration to approximate $\displaystyle\int_2^5 \log_3 x\ dx$.

24.   Let $s(x) = .5[e^x - e^{-x}]$ and $c(x) = .5[e^x + e^{-x}]$.

      (a)   Draw both graphs on one coordinate system.
      (b)   Check to see if the functions are even or odd.
      (c)   Draw the graphs of $y = c(x) + s(x)$ and $y = c(x) - s(x)$

25.   Use numerical integration to approximate the integrals

$$\int_2^3 \ln\sqrt{x^3 + x + 1}\ dx \qquad \text{and} \qquad \int_2^3 \ln(x^3 + x + 1)\ dx$$

How are they related?   Why?

26.   Approximate the roots of
$$x^2 e^{3x} - x^3 \cos(x) + 57\ln(2x+3) = 0$$

27.   Determine any maxima or minima for
$$y = e^{-2x} + \ln(x^2 + 5) + x^3$$

28.   Determine the concavity of $y = \ln(x^2 - e^{\cos x} + 3)$ for $-2 \le x \le 2$

14

HYPERBOLIC FUNCTIONS

The hyperbolic functions behave much like the trigonometric functions and therefore have names much like them. The hyperbolic sine, denoted by sinh, is the odd part of the natural exponential function e^x and the hyperbolic cosine, denoted by cosh, is the even part of e^x. Recall that this means

$$\sinh x = \frac{e^x - e^{-x}}{2} \quad \text{and} \quad \cosh x = \frac{e^x + e^{-x}}{2}$$

The other four hyperbolic functions are defined using the analogy to trigonometric functions.

$$\tanh x = \frac{\sinh x}{\cosh x} \quad \coth x = \frac{1}{\tanh x} \quad \text{sech } x = \frac{1}{\cosh x} \quad \text{csch } x = \frac{1}{\sinh x}$$

With the TI-81, the graphs of the three basic hyperbolic functions are drawn by choosing Y_1 (or one of the other Y variables), using the MATH key, choosing HYP, and designating the function. The screen appears as in Figure 14-1.

FIGURE 14-1

```
MATH    NUM    HYP    PRB
1:sinh
2:cosh
3:tanh
4:sinh-1
5:cosh-1
6:tanh-1
```

The graphs of the three basic hyperbolic functions are given in Figure 14-2

FIGURE 14-2

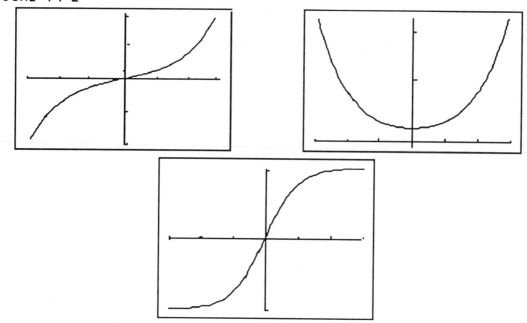

As with trigonometric functions, we cannot expect all hyperbolic functions to have complete inverses but inverses do exist for two of the basic functions and a partial inverse exists for the third. The graphs are given in Figure 14-3.

FIGURE 14-3

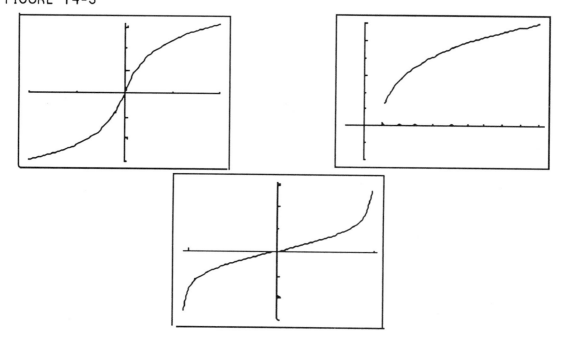

Your calculator can be used to draw graphs of rather complicated functions involving hyperbolic functions.

Example 1. The graph of
$$y = (x\sinh x)^2 - (\cosh x)^3$$
is given in Figure 14-4.

FIGURE 14-4

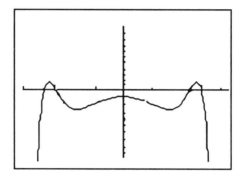

Identities involving hyperbolic functions can be tested as were those involving trigonometric functions.

Example 2. Are the functions $\cosh^2 x$ and $\sinh^2 x$ related by an identity?

SOLUTION: Let Y_1, Y_2, and Y_3 be as in Figure 14-5 with the three graphs given in Figure 14-6.

FIGURE 14-5

> ₁Y₁=(cosh X)2
> ₁Y₂=(sinh X)2
> ₁Y₃=Y₁-Y₂
> ₁Y₄=

FIGURE 14-6

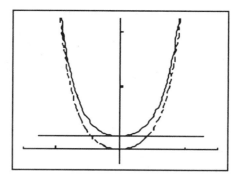

From the graphs, we might expect that
$$(\cosh x)^2 - (\sinh x)^2 = 1.$$
The equation holds and can be verified by algebraic means. ‖

Hyperbolic functions have a number of applications in physics and engineering. One of these involves a curve called a *catenary*.

A catenary: Suppose a cable with uniform density is suspended between two points of equal height and has no other weight hanging on it. If an x,y coordinate system is placed so that the x-axis is parallel to ground level and the y-axis passes through the low point on the curve, then the shape of the cable is the shape of the graph of
$$y = \frac{c}{k} \cosh \frac{kx}{c}$$
where k is the density of the cable, c is the horizontal tension on the cable at its lowest point and the low point is a distance of $\frac{c}{k}$ above the x-axis.

Example 3. The graph given in Figure 14-7 is a catenary with c = 5 and k = 2.
$$y = \frac{5}{2} \cosh \frac{2x}{5}$$

FIGURE 14-7

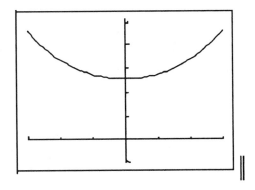

The techniques discussed in earlier sections can be applied to expressions involving hyperbolic functions.

Example 4. Suppose a cable is hanging between supports and has the shape of the graph of

$$y = \frac{250}{13} \cosh \frac{13x}{250}$$

Determine an approximation to the slope of the cable when x = 15.

SOLUTION: The graph of the equation is given in Figure 14-8.

FIGURE 14-8

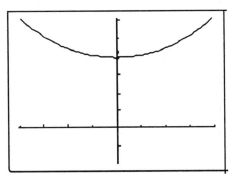

The slope of the tangent line can be approximated by the numerical differentiation method. The results are displayed in Figure 14-9. The cable has a slope of approximately 0.86 at x = 15.

FIGURE 14-9

```
15→X
                    15
NDeriv(1250/13)c
osh (13X/250),,D
1)
        .8615331665
```

Example 5. Determine the concavity of
$$y = x^2\sinh x + 4\cosh x$$

SOLUTION: The second derivative is
$$y'' = x^2\sinh x + 4x\cosh x + 2\sinh x + 4\cosh x$$

Both graphs are drawn in Figure 14-10.

FIGURE 14-10

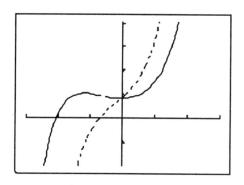

The second derivative seems to have just one zero between -2 and 0. The results of applying ZERO are displayed in Figure 14-11.

FIGURE 14-11

```
?0
ZERO AT
        -.6523731803
Y₁=
        -1.633E-9
Y₂=
    4.584039841
```

The graph is concave downward on $(-\infty, -.6523731803]$ and concave upward on $[-.6523731803, \infty)$.

EXERCISES

1. Draw the graphs of
$$y = \coth x, \quad y = \operatorname{sech} x, \quad \text{and} \quad y = \operatorname{csch} x$$

2. Draw the graphs of
$$y = \operatorname{csch}^{-1} x \qquad y = \operatorname{sech}^{-1} x \qquad y = \coth^{-1} x$$

3. Draw the graph of
$$y = \sinh\left[\frac{x}{x^2 + 1}\right]$$

4. Draw the graph of
$$y = \operatorname{csch}(\sin x)$$

5. Draw the graph of
$$y = \sin(\cosh x)$$

6. Draw the graph of
$$y = \ln(\cosh x)$$

7. Draw the graph of
$$y = \ln(\tanh x)$$

8. Draw the graph of
$$y = \cosh x - \sinh x$$

Does the graph look familiar?   What does it look like?

9. Draw the graphs of
$$Y_1 = \cosh x + \sinh x \qquad \text{and} \qquad Y_2 = e^x$$

Does there appear to be an identity here?   Does this give any insight to Exercise 8 ?

10. Draw the graphs of
$$Y_1 = \operatorname{sech}^2 x, \ Y_2 = \tanh^2 x, \ \text{and} \ Y_3 = Y_1 + Y_2$$

Does there appear to be an identity here?

11. Draw the graphs of

$$Y_1 = \cosh \frac{x}{2} \quad\quad \text{and} \quad\quad Y_2 = \sqrt{\frac{\cosh x + 1}{2}}$$

Does there appear to be an identity here?

12. Draw the graphs of

$$Y_1 = \sinh^{-1} x \quad\quad \text{and} \quad\quad Y_2 = \ln\left[x + \sqrt{x^2 + 1}\right]$$

Does there appear to be an identity here?

13. Compute an approximation to f'(1.4) if

$$f(x) = \frac{\cosh x + 1}{\tanh^2 x + 3}$$

14. Compute an approximation to g'(2.1) if

$$g(x) = \cosh\left[\frac{2 + \sinh x}{x^2 + 1}\right]$$

15. Is the hyperbolic tangent even, odd, or neither?   How about coth x ?

16. Is the hyperbolic secant even, odd, or neither?   How about csch x ?

17. Determine any local maxima and minima for

$$y = x\sinh x + 2\cosh x - x$$

18. Use the Trapezoidal Rule to approximate

$$\int_1^3 x^2 \cosh x \, dx$$

19. Use Simpson's Rule to approximate

$$\int_{-1}^2 \sinh(\cosh(x)) \, dx$$

20. Determine the concavity of

$$y = \frac{\sinh x}{x^2 + 1}$$

1 5

NUMERICAL INTEGRATION: APPLICATIONS

Many of the definite integrals that arise in applications of calculus cannot be computed using the Fundamental Theorem of Calculus because the integrands do not have elementary antiderivatives. This does not mean that the techniques cannot be used; it just means that an approximation is the best we can hope for in computing the solution. Those who have access to a computer with sophisticated software can compute very good approximations. However, the methods developed in Section 11 provide sufficient approximations for many such problems.

PROBABILITY:

A function $p(x)$ such that

$$1. \quad p(x) \geq 0 \text{ for all } x$$

$$2. \quad \int_{-\infty}^{\infty} p(x) \, dx = 1$$

can be viewed as a probability density function with the probability that $a \leq x \leq b$ given by

$$\int_a^b p(x) \, dx$$

One of the most important such functions is the normal density function

$$p(x) = \frac{1}{\sqrt{2\pi}} e^{-x^2/2}$$

Example 1. Approximate the probability that $-1 \leq x \leq 2$ with the normal density function.

SOLUTION: We need to compute

$$\int_{-1}^{2} \frac{1}{\sqrt{2\pi}} e^{-x^2/2} \, dx$$

The number needed is the area of the region shaded in Figure 15-1

FIGURE 15-1

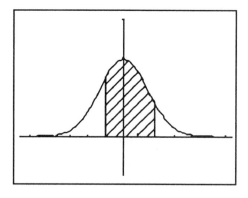

There is no simple antiderivative for this function. In this case we will use Simpson's Rule for the approximation. The results are displayed in Figure 15-2.

FIGURE 15-2

```
?-1
UPPER LIMIT
?2
N DIVISIONS
?256
INTEGRAL-
        .8185946141
```

The probability is approximately 0.8186. ‖

If you are going to do quite a bit of work with normal density functions or other common probability functions, then you should expect to use tables or computer programs.

ARC LENGTH:

The length of the graph of y = f(x) from x = a to x = b is given by

$$L = \int_a^b \sqrt{1+[f'(x)]^2}\ dx$$

Frequently this formula presents us with an integral that we cannot compute with simple methods.

Example 2. Compute the length of the curve $y = \sin(2x)$ from $x = 0$ to $x = \pi$.

SOLUTION: The graph of the curve is given in Figure 15-3.

FIGURE 15-3

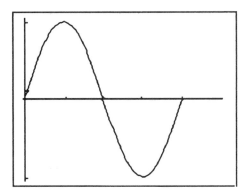

In this case, $y' = 2\cos(2x)$ and the length is given by

$$L = \int_0^\pi \sqrt{1 + 4\cos^2(2x)}\ dx$$

No antiderivative is readily available but we can approximate the length with Simpson's Rule.   In this case we use 64 subdivisions. The results are given in Figure 15-4.

FIGURE 15-4

```
?0
UPPER LIMIT
?π
N DIVISIONS
?64
INTEGRAL IS
          5.270367163
```

The length is about 5.27. ‖

VOLUME OF A SOLID OF REVOLUTION:

The volume of a solid of revolution generated by revolving the region bounded by the x-axis, the lines $x = a$ and $x = b$ with $a < b$, and the graph of the non-negative function $f(x)$ about the x-axis is given by

$$V = \int_a^b \pi[f(x)]^2\ dx$$

Example 3. Determine the volume of the solid generated by revolving the region bounded by the x-axis, the lines x = 2 and x = 5, and the curve

$$y = \frac{1}{\sin^2(x) + 3}$$

about the x-axis.

SOLUTION: The volume is given by

$$V = \int_{2}^{5} \pi \left(\frac{1}{\sin^2(x) + 3} \right)^2 dx$$

The curve is given in Figure 15-5 with RANGE [2,5,1,.1,.4,1.1].

FIGURE 15-5

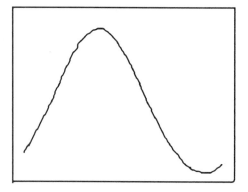

We know of no simple antiderivative for the integrand but we can apply the Trapezoidal Rule with 128 subdivisions to obtain an approximation. The results are displayed in Figure 15-6.

FIGURE 15-6

```
?2
ENTER B
?5
?128
APPROX IS
        .8013710426
```

The volume is about 0.801. ‖

SURFACE AREA OF A SOLID OF REVOLUTION:

The area of the surface of revolution generated by revolving the region bounded by the x-axis, the lines $x = a$ and $x = b$ with $a < b$, and the graph of a non-negative function $y = f(x)$ about the x-axis is given by

$$S = 2\pi \int_a^b f(x)\sqrt{1 + [f'(x)]^2}\,dx$$

Example 4. Compute the surface area of the solid of revolution obtained by revolving the region bounded by the x-axis, the graph of the curve
$$y = 4 + \sinh(x),$$
and the lines $x = 0$ and $x = 3$ about the x-axis.

SOLUTION: The graph of the curve is given in Figure 15-7.

FIGURE 15-7.

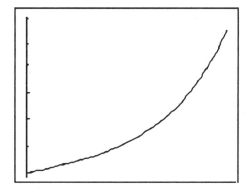

The surface area of the solid of revolution is given by

$$S = 2\pi \int_0^3 (4 + \sinh(x))\sqrt{1 + (\cosh(x))^2}\,dx$$

We approximate with the Trapezoidal Rule and 128 subdivisions; the results are displayed in Figure 15-8.

FIGURE 15-8

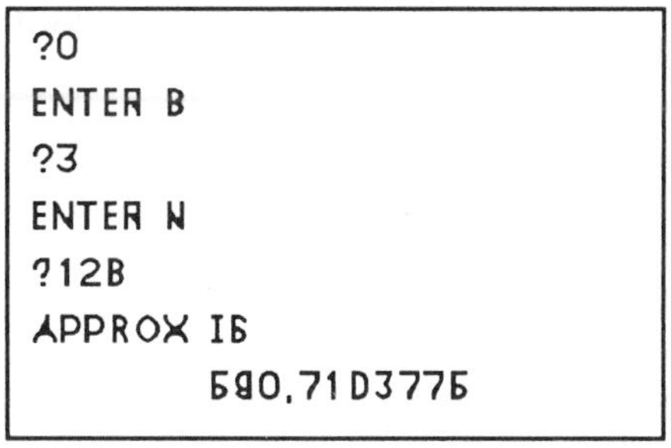

```
?0
ENTER B
?3
ENTER N
?12B
APPROX IƂ
        Ƃ90.71D377Ƃ
```

The area is approximately 590.7. ‖

Your calculus book contains many problems that can be solved using definite integration. We urge you to apply the methods of this section to some of the exercises in your text.

EXERCISES

1. Estimate the probability that $1 \leq x \leq 7$ with the standard normal probability density function.

2. Estimate the probability that $-1 \leq x \leq 1$ with the standard normal probability density function.

3. Approximate the length of the graph of $y = x + \sin(\pi x)$ from $x = 0$ to $x = 4$.

4. Approximate the length of the curve $y = \dfrac{4}{x^2 + 5}$ from $x = -1$ to $x = 3$.

5. Determine the volume of the solid of revolution generated by revolving the region bounded by the curve $y = \ln(4 + e^x)$, the lines $x = 1$ and $x = 4$, and the x-axis about the x-axis.

6. Determine the volume of the solid of revolution generated by revolving the region bounded by the curve $y = \sin^2(e^x)$, the lines $x = -3$ and $x = 4$, and the x-axis about the x-axis.

7. Compute the surface area of the solid of revolution in Exercise 5.

8. Compute the surface area of the solid of revolution in Exercise 6.

9. Compute the volume of the solid generated by revolving the portion of the curve $y = 4 + 3x + 9x^2 - x^4$ above the x-axis about the x-axis. (Hint: First use **ZERO** to find the limits of integration.)

10. Compute the volume of the solid of revolution generated by revolving the region bounded by $y = x^3 + 4x^2 + 3x + 2$ and the line $y = 5x + 1$ about the x-axis. (Hint: Use **ZERO**.)

11. The work done by a variable force $F(x)$ as it moves an object in a straight path from $x = a$ to $x = b$ is given by $\int_a^b F(x)\ dx$. Determine the amount of work done by a force $F(x) = \cos^2(\ln(x^2 + 8) + e^{4x})$ as it moves an object from $x = 0$ to $x = 10$.

12. Compute the work done by a force $F(x) = e^{\sin x}$ as it moves an object from $x = 1$ to $x = 5$.

13. The volume of a solid of revolution can also be given using the method of cylindrical shells. If the region bounded by the x-axis, the lines $x = a$ and $x = b$ with $0 \le a < b$ and the graphs of $f(x)$ and $g(x)$ with $g(x) \le f(x)$ for $a \le x \le b$ is revolved about the y-axis, the volume is given by

$$\int_a^b 2\pi x[f(x) - g(x)]\ dx$$

Determine the volume of the solid of revolution generated by revolving the region bounded by $x = 2$, $x = 4$, and the graphs of $f(x) = e^{x^3}$ and $g(x) = \ln(x^2 - 1)$ about the y-axis.

14. Determine the volume of the solid of revolution generated by revolving the region bounded by $y = \sqrt{x^3 + 5x^2 + 2x + 7}$, $y = \cos^2(x^4 + 3)$, the lines $x = 1$ and $x = 4$ about the y-axis.

15. Determine the volume of the solid of revolution generated by revolving the region bounded by the x-axis and the curves $y = e^{x^2 - 3}$ and $y = 3\cos x$ about the y-axis.

16.   Suppose that the velocity of an object moving in a linear path is given by $v(t) = t^3[\sin(5t + 7) + e^t]$. Approximate the total distance traveled by the object from time $t = 2$ to time $t = 7$.

17.   Suppose the daily costs of a business are given by
$$c(t) = 80\cos^4(.001t) + 31\ln(t^3 + t^4) + 17e^{3-t^{.6}}.$$
What are the total costs from day 5 through day 2 3?

EXTENSION

18.   We know that $\displaystyle\int_0^4 \frac{\sin x}{x}\, dx$ is an improper integral.   Use the Trapezoidal

Rule to approximate it.

19.   We know that $\displaystyle\int_0^1 \ln(1-x)\, dx$ is an improper integral.   Use the Midpoint

Rule to approximate it.

20.   Determine the volume of the solid of revolution generated by revolving the region bounded by $x = 1$, the y-axis, and $y = x^{-1/3}$ about the x-axis. Notice that this is an improper integral.

16

SEQUENCES AND SERIES

In ordinary language, the word "sequence" means a collection of objects that are ordered, one after another. This concept is made precise in mathematics by defining a sequence to be a special type of function.

DEFINITION:

A sequence is a function with the set of positive integers as its domain.

The function values for a sequence are usually denoted by symbols such as

$$a_1, a_2, a_3, \ldots, a_n, \ldots$$

rather than

$$f(1), f(2), f(3), \ldots, f(n), \ldots$$

and are called the *terms* of the sequence. A sequence is commonly denoted by a symbol such as $\{a_n\}$ or a formula in n that tells us what the value is for any positive integer n. Expressions such as

$$\{n/(n+1)\}, \quad \{e^{-n}\}, \quad \text{and} \quad \{\sin(n)\}$$

are examples of such descriptions.

The sequences that are of the most interest and importance to us are those in which the terms approach some number as n gets large. The precise definition of the limit of a sequence is much like the definition of the limit of a function $f(x)$ as $x \to \infty$ and we use some of the same notation.

DEFINITION:

If L is a number and $\{a_n\}$ is a sequence,

$$\lim_{n \to \infty} a_n = L$$

if for each $\varepsilon > 0$, there exists $M > 0$ such that

$$|a_n - L| < \varepsilon \text{ whenever } n > M.$$

Sequences that have a limit are said to *converge* ; those that do not have a limit are said to *diverge*.

When we approximated limits of functions in earlier sections, we actually were finding limits of sequences of function values. We chose a sequence of numbers that was converging to the number where we were computing the limit.   Then we evaluated the function at these numbers and the limit was the number to which these function values were converging. The primary difference is that now we restrict the values of the variable to positive integers.

The problem of finding limits of sequences is closely related to finding horizontal asymptotes; we are looking for numbers that the function or sequence gets close to as the variable x or n gets large.   Graphic techniques for finding horizontal asymptotes can sometimes be used for sequences.

Example 1. Determine the limit of the sequence
$$\{(8n+13)/(2n+7)\}$$

SOLUTION: We will apply both a graphics technique and the program **LIMIT** developed in an earlier section.

First of all, let
$$Y_1 = (8X-13)/(2X+7)$$
and draw the graph in Figure 16-1 with a RANGE of $[-100,1000,100,-1,10,2,1]$.

FIGURE 16-1

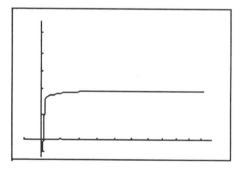

The graph seems to be getting close to **4** and we would expect that **4** is the limit.

We can use $LIMIT$ to solve the problem by letting
$$Y_1 = (8X^{-1} - 13)/(2X^{-1} + 7)$$
and computing the right limit as $X \to 0^+$. The results are displayed in Figure 16-2.

FIGURE 16-2

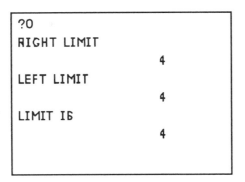

```
?0
RIGHT LIMIT
                    4
LEFT LIMIT
                    4
LIMIT IS
                    4
```

The limit is 4. ▌

When we are approximating limits of sequences, we are often better served by programs designed for that purpose. The earlier programs do not examine function values for as large values of x as we may need.

The program $LIMSEQ$ is given below and can be stored in your calculator. It does not compute the function values for all values of n up to 2^{30} but does compute them for $n = 1,2,4,8,16,...,2^{30}$, a total of 31 values. As soon as the terms of the sequence are within 2^{-25} of each other, the program will give the last term computed as the limit. This is not completely accurate but we are only expecting an approximation of the limit.

```
Prgm :LIMSEQ                    :Goto 8
:12345→B                          :Goto 1
:2^(-25)→E                       :Lbl 7
:Disp"ENTER N"                   :Disp"LIMIT IS"
:1→X                             :Disp A
:Lbl 1                           :Goto 9
:Y₁→A                            :Lbl 8
:If abs(A-B)<E                   :Disp"AFTER 2^3
:Goto 7                          0 TERMS NO LIMIT
:A→B                             :LBL 9
:2*X→X                           :END
:If X>2^ 30
```

If you want a different degree of accuracy than given, edit the program and make the appropriate changes. As before, we first list Y_1 as the general term of the sequence using X for the variable instead of n.

Example 2. Approximate the limit of
$$\{(5n + 3)/(2n + 1)\}.$$

SOLUTION: Express the general term of the sequence.
$$Y_1 = (5X+3)/(2X+1)$$
The results of applying **LIMSEQ** are given in Figure 16-3.

FIGURE 16-3

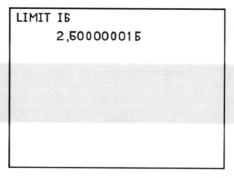

```
LIMIT IS
    2.500000015
```

The limit is 2.5. ‖

This sequence presented no challenge; we could have done the problem easily without the calculator. Limits of more complicated sequences can be found with the program and would be difficult otherwise.

Example 3. Approximate the limit of
$$\{(5n^2 + \ln(n))/\sqrt{n^4+7}\}.$$

SOLUTION: We list the general term as
$$Y_1 = (5X^2 + \ln X)/(\sqrt{X^4 + 7})$$
and apply **LIMSEQ.** The results are given in Figure 16-4.

FIGURE 16-4

```
LIMIT IS
     5.00000001
```

The limit is 5. ‖

The calculator can also be used to determine when a sequence does not have a limit, at least as far as 2^{30} terms.

Example 4. Approximate the limit of
$$\{(\sin n)/(2+\cos n)\}.$$

SOLUTION: We let
$$Y_1 = (\sin X)/(2 + \cos X)$$
and apply **LIMSEQ.** The results are given in Figure 16-5.

FIGURE 16-5

```
AFTER 2^30 TERMS
NO LIMIT
```

This sequence diverges. ‖

Remember that in using these programs, you cannot expect the calculator to compute beyond its capacity. Large powers of large numbers may lead to an error signal.

INFINITE SERIES

DEFINITION:

An infinite series is an indicated infinite sum

$$\sum_{n=1}^{\infty} a_n = a_1 + a_2 + a_3 + \cdots + a_n + \cdots$$

It is said to converge if the sequence of partial sums

$$S_n = a_1 + a_2 + a_3 + \cdots + a_n$$

converges and to diverge otherwise.

NOTATIONAL CONVENTION:

We will eliminate the subscripts and superscripts and just write

$$\sum a_n$$

There are many tests for convergence of series; we list one here and give an example of its use.

THE RATIO TEST:

If $\sum a_n$ is an infinite series such that

$$\lim_{n \to \infty} \left| (a_{n+1} / a_n) \right| < 1$$

then the series converges. If the limit of the ratio is greater than 1, then the series diverges. If the limit is 1 or does not exist, the test is not applicable to this series.

Example 5. Use the ratio test to determine if the series

$$\sum \frac{1}{10^n}$$

converges.

SOLUTION: The ratio is

$$Y_1 = (1/10^{X+1})/(1/10^X)$$

The results using **LIMSEQ** are given in Figure 16-6.

FIGURE 16-6

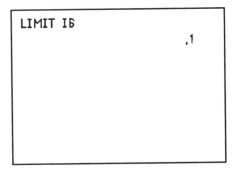

The limit of the ratio is .1 and the series converges. ‖

Example 6. Use the ratio test to determine if the series
$$\Sigma \ e^{1/n}$$
converges.

SOLUTION: Let
$$Y_1 = \ e^{1/(X+1)}/e^{1/X}$$
and apply **LIMSEQ.** The results are given in Figure 16-7.

FIGURE 16-7

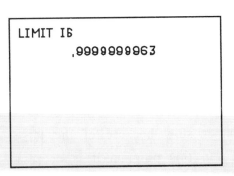

In this case, the limit is 1 and the ratio test fails; it does not tell us whether or not the series converges. Actually,
$$\lim_{n \to \infty} \ e^{1/n} = 1$$
and the series diverges. ‖

In using tests for the convergence of a series that involve convergence of an associated sequence, remember that you must stay within the computing capacity of your calculator. For example, the programs usually cannot be used for sequences involving $n!$ because of the magnitude of the numbers. They may be applicable if you do some algebraic simplification first.

The tests that determine whether or not a series converges usually do little to approximate its limit. The program SERIES computes the sum of the first 2^{10} terms of the sequence, displays this sum and indicates whether or not the sequence of sums appears to have a limit. You can store it in your calculator. Note that this program assumes that the first term is a_1, not a_0.

```
Prgm: SERIES
:0→S
:.00000001→E
:1→X
:Lbl 1
:S+Y₁→K
:If abs(S-K)<E
:Goto 7
:K→S
:1+X→X
:If X>2^10
:Goto 8
:Goto 1
:Lbl 7
:Disp"SUM IS"
:Disp S
:Goto 9
:Lbl 8
:Disp"AFTER 2^1
0 TERMS, NO LI
MIT"
:DISP"Y₁="
:DISP Y₁
:DISP "S="
:DISP S
:Lbl 9
:End
```

The program is rather slow because of the large number of computations to be done. In this program it is necessary to compute and sum 2^{10} terms whereas LIMSEQ only considers 30 terms.

Example 7. Approximate the limit of the series from Example 5.

SOLUTION; We know that the series converges.
Let $Y_1 = 1/10^X$
The results are given in Figure 16-8.

FIGURE 16-8

```
SUM IS
      .1111111
```

The series converges to approximately 0.1111111. ‖

We know that some series converge because they are dominated by series that are known to converge. We can use **SERIES** to approximate the limit.

Example 8. Approximate the sum of the series
$$\Sigma\ (\sin n)/3^n$$

SOLUTION: We know that the series converges because it is dominated by
$$\Sigma\ 1/3^n$$

which is known to be a convergent geometric series.
Let $Y_1 = (\sin X)/3^X$

The results are displayed in Figure 16-9

FIGURE 16-9

```
SUM IS
      .37353440655
```

The sum is approximately 0.37353. ‖

As with sequences, remember that you cannot expect results from your calculator that are beyond its capacity. In some cases, some algebraic simplification will enable you to use the calculator.

The programs here are only a sample of the technology available for dealing with sequences and series and, in fact, are rather simplistic and very limited. If you do a great deal of work with sequences or series, you should make use of some of the computer software that is available.

EXERCISES

1. Determine whether or not each of the following sequences converge. If so, find the limit.
   (a) $\{(7n+2)/(5n+2)\}$
   (b) $\{(2n^2 + 5)/(3n^3 - 7n + 9)\}$
   (c) $\{(5n^3 + 4n - 1)/(3n^2 + 2)\}$
   (d) $\{(\ln(x^{3.7} + 3))/x^2\}$

2. Determine whether or not each of the following sequences converge. If so, find the limit.
   (a) $\{e^{\sin(n)}\}$
   (b) $\{\sin(n^{-1})\}$
   (c) $\{\cos(n^{-2})\}$
   (d) $\{n^{\sin(n)}\}$

3. Determine whether or not each of the following sequences converge. If so, find the limit.
   (a) $\{(1 + 1/n)^n\}$
   (b) $\{(4 - 3/n)/(\pi - 2/n)\}$
   (c) $\{\sin(n^2)/(n^3 + 1)\}$
   (d) $\{\sin(n)/\cos(3n)\}$

4. Determine whether or not each of the following sequences converge. If so, find the limit.
   (a) $\{\ln(2+\sin n)\}$
   (b) $\{n^2 e^{-n}\}$
   (c) $\{\ln(e^{-n} + n^3)\}$
   (d) $\{\sin(e^{(n+5)/n})\}$

5. Approximate the sum of the series $\Sigma \ (1/5)^n$

6. Approximate the sum of the series $\Sigma \; (\cos \; n)/n^2$.

7. Approximate the sum of the series $\Sigma \; e^{-n}$.

8. Approximate the sum of the series $\Sigma \; \cos(n)/2^n$.

9. Does the series $\Sigma \; 4/7^x$ converge? If so, what is the limit?

10. Does the series $\Sigma \; e^{n/(n+1)}$ converge? If so, what is its limit?

11. Use a graphic technique to determine if $\{e^{\sin(\pi n)}\}$ converges. Now use **LIMSEQ**. Are your answers different? Explain the results.

12. Choose an alternating sequence $\{a_n\}$ that converges to 0. Now construct the series $\Sigma \; a_n$. Must this series converge? Why? Approximate its limit.

17

TAYLOR POLYNOMIALS

Polynomial functions are the simplest to deal with in differentiation, integration, and evaluation. Taylor polynomials provide a way of approximating many common and useful functions by polynomial functions.

Example 1. The graphs of
$$y = \sin x \quad \text{and} \quad y = x - x^3/6 + x^5/120$$
are given in Figure 17-1 and are drawn with a RANGE setting of $[-\pi, \pi, 1, -2, 2, 1, 1]$.

FIGURE 17-1

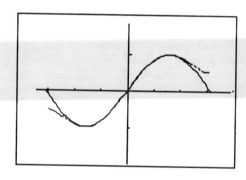

The graphs seem to be the same from -2 to 2. By evaluating both functions, we see in Figure 17-2 that they differ by less than .01 for x = 1.

FIGURE 17-2

1→K	
	I
Y₁	
	.841470984B
Y₂	
	.834165972B

DEFINITIONS: If f is a function with n derivatives at x = a, the polynomial

$$P_n(x) = f(a) + f'(a)(x-a) + \frac{f''(a)}{2}(x-a)^2 + \frac{f'''(a)}{6}(x-a)^3 + \frac{f^{(4)}(a)}{24}(x-a)^4 +$$

$$\cdots + \frac{f^{(n)}(a)}{n!(x-a)^n} \qquad \text{where } n! = 1 \cdot 2 \cdot 3 \cdots n$$

is called the *nth degree Taylor polynomial* for f at a.

If a = 0, then P_n is called the nth degree *Maclaurin* polynomial

for f.

The function

$$R_n(x) = f(x) - P_n(x)$$

is called the *nth degree Taylor remainder* for f(x).

For many functions,

$$\lim_{n \to \infty} R_n(x) = 0$$

which means that such functions can be approximated quite well by their Taylor polynomials.

Example 2. Determine the 4th degree Maclaurin polynomial for y = cos x and draw the graphs of the polynomial and the function on the same coordinate system.

SOLUTION: First we need to determine the values of the derivatives at 0.

Derivatives	Values at 0
y = cos x	1
y' = -sin x	0
y" = -cos x	-1
y'''= sin x	0
$y^{(4)}$ = cos x	1

The 4th degree Maclaurin polynomial is

$$P_4(x) = 1 - \frac{1}{2}x^2 + \frac{1}{24}x^4$$

The graphs are given in Figure 17-3

FIGURE 17-3

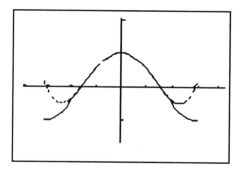

Example 3. Determine the 5th degree Taylor polynomial at 1 for

$$y = \ln x$$

and draw the graphs of the function and the polynomial on [.5,2.5].

SOLUTION:

Derivatives	Values at 1
$y = \ln x$	0
$y' = x^{-1}$	1
$y'' = -x^{-2}$	-1
$y''' = 2x^{-3}$	2
$y^{(4)} = -6x^{-4}$	-6
$y^{(5)} = 24x^{-5}$	24

The polynomial is

$$P_5 = (x-1) - \frac{1}{2}(x-1)^2 + \frac{2}{6}(x-1)^3 - \frac{6}{24}(x-1)^4 + \frac{24}{120}(x-1)^5$$

or

$$P_5 = (x-1) - \frac{1}{2}(x-1)^2 + \frac{1}{3}(x-1)^3 - \frac{1}{4}(x-1)^4 + \frac{1}{5}(x-1)^5$$

The graphs are given in Figure 17-4 with a RANGE setting of [.5,2.5,1,-1,2,1,1]

FIGURE 17-4

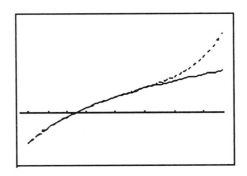

The internal programs that some calculators use to evaluate trigonometric, logarithmic, exponential and many other functions are based on the Taylor polynomials for these functions. It is truly amazing that Taylor and Maclaurin developed these ideas more than two hundred years before the invention of the computing instruments that would put them into common use.

It is natural to wonder just how good these polynomial approximations really are. The following theorem answers this question for many functions.

TAYLOR'S THEOREM: If f has n+1 derivatives in an interval I containing c, then for each x in I, there is a number z between x and c such that

$$f(x) = P_n(x) + R_n(x) \quad \text{where} \quad R_n(x) = \frac{f^{(n+1)}(z)}{(n+1)!}(x - c)^{n+1}$$

The chances of finding such a z are slight but for many functions we can find an upper bound for $|f^{(n+1)}|$ on the interval I and this will suffice.

Example 4. Show that the fourth degree Maclaurin polynomial provides an approximation for $\cos x$ that is correct to within .002 for $-1 < x < 1$.

SOLUTION: From Example 2, we know that

$$P_4 = x = \frac{1}{2}x^2 + \frac{1}{24}x^4 \quad \text{with} \quad y^{(5)} = -\sin x$$

For any number z between x and 0,

$$-1 < z < 1 \quad \text{and} \quad \left|y^{(5)}(z)\right| = \left|-\sin z\right| < 1$$

Also $\left|x^6\right| < 1$ since $|x| < 1$

Therefore $\qquad \left| R_n(x) \right| = \left| \dfrac{-\sin z}{6!} x^6 \right| < \dfrac{1}{6!} = \dfrac{1}{720} \approx .00138$

Since $f(x) = P_4(x) + R_n(x)$, $P(x)$ differs from $\cos x$ by less than $.00138$ and thus by less than $.002.$ ‖

We can also use graphic techniques to determine bounds for R_n.

Example 5. Determine the sixth degree Maclaurin polynomial for $y = \sinh x$ for values of x between -1 and 1 and estimate its accuracy.

SOLUTION:

Derivatives	Values at 0
$f(x) = \sinh x$	$f(0) = 0$
$f'(x) = \cosh x$	$f'(0) = 1$
$f''(x) = \sinh x$	$f''(0) = 0$
$f'''(x) = \cosh x$	$f'''(0) = 1$
$f^{(4)}(x) = \sinh x$	$f^{(4)}(0) = 0$
$f^{(5)}(x) = \cosh x$	$f^{(5)}(0) = 1$
$f^{(6)}(x) = \sinh x$	$f^{(6)}(0) = 0$
$f^{(7)}(x) = \cosh x$	

The polynomial is
$$P_6(x) = x + \frac{x^3}{3!} + \frac{x^5}{5!}$$

The remainder term is
$$\left| R_6(z) \right| = \left| \frac{\cosh z}{7!} x^7 \right| \le \frac{\cosh z}{7!} \quad \text{since } |x| \le 1$$

The graph of $Y_1 = \dfrac{\cosh x}{7!}$ is given in Figure 17-5 with a RANGE setting of $[-1,1,1,-.001,.001,.001,1]$.

FIGURE 17-5

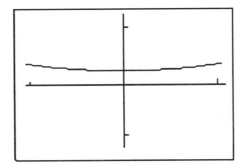

The y values are between 0 and .001 and the sixth degree polynomial approximation is accurate to within .001 on [-1,1]. ‖

EXERCISES

1. Determine the sixth degree Maclaurin polynomial for cosx and draw the graphs of the function and the polynomial with the Trig RANGE setting.

2. Determine the fourth degree Maclaurin polynomial for coshx and draw the graphs of the function and the polynomial for $-2 \leq x \leq 2$. Use the polynomial to approximate cosh(.5).

3. Determine the fifth degree Maclaurin polynomial for $y = e^x$. Draw both graphs for $-1 \leq x \leq 1$. Use the polynomial to approximate $e^{.2}$

4. Determine the third degree Maclaurin polynomial for $y = \tan^{-1}x$. Draw both graphs for $-3 \leq x \leq 3$.

5. Determine the fourth degree Taylor polynomial for $y = \tan x$ at $a = \pi/4$. Draw both graphs for $0 < x < \pi/2$. Use the polynomial to approximate tan(1). Use graphic techniques to extimate the accuracy.

6. Determine the third degree Maclaurin polynomial for $y = \sin x$. Draw the graphs of both with the Trig RANGE setting. Use the graph of the polynomial to approximate $\sin(.4)$. Use graphic techniques to estimate the accuracy.

7. Determine the fifth degree Taylor polynomial for $y = \dfrac{1}{x}$ at $a = 1$. Draw both graphs for $.5 \leq x \leq 1.5$. Determine a bound for R_5 on the interval and use it to estimate the accuracy of the approximation.

8. Determine the fourth degree Taylor polynomial for $y = \csc x$ at $x = \pi/4$. Draw both graphs for $0 < x < \pi/2$. How good is the approximation on $[\pi/8, 3\pi/8]$??

9. Determine the fifth degree Maclaurin polynomial for $y = e^{-x}$. Draw both graphs for $-1 \leq x \leq 1$. Determine a bound for the approximation.

10. Determine the fifth degree Maclaurin polynomial for $y = \sinh 3x$. Draw both graphs for $-2 \leq x \leq 2$. Determine a bound for the approximation on $[-1,1]$..

11. Determine the sixth degree Maclaurin polynomial for
$$y = 2x^5 - 10x^4 + 3x^3 + 2x^2 - 3x + 1.$$
What can you say about it?

12. Determine the fifth degree Taylor polynomial for
$$y = x^4 - 4x^3 - 6x^2 + 2x + 3$$
at $a = 1$. Draw its graph and the graph of the given function for $-4 \leq x \leq 4$. Explain your result.

13. Differentiate the polynomial obtained in Example 2 and compare it to the answer to Exercise 6. Explain your result.

14. Differentiate the answer to Exercise 3 and compare it to the answer to Exercise 3. Explain your result.

15. Use your answers to Exercises 3 and 9 to determine the fifth degree Maclaurin polynomial for $y = \sinh x$.

16. Determine the fifth degree Maclaurin polynomial for
$$y = e^x + x^2 - x.$$
Explain how this polynomial is related to your answer to Exercise 3.

18

CONIC SECTIONS

If a plane intersects a right circular cone in a curve, the curve is called a *conic section*. If the plane passes through the cone, the curve is called an *ellipse*; if the plane is parallel to one side of the cone, the curve is a *parabola*, and if the plane intersects both nappes of the cone, the curve is called a *hyperbola*. All three possibilities are shown in Figure 18-1.

FIGURE 18-1

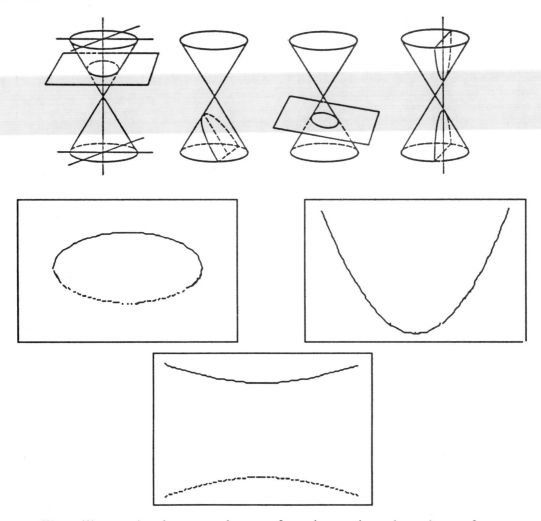

We will not develop any theory of conic sections here but refer you to any precalculus or calculus book for the definitions of axes, vertices, centers, and for basic properties.

Recall that if the axis of a parabola is parallel to the y-axis, the parabola has an equation of the form

$$y = ax^2 + bx + c$$

Similarly, if the axis of the parabola is parallel to the x-axis, the parabola has an equation of the form

$$x = ay^2 + by + c$$

The graphs of

$$y = 2x^2 - 3x + 5 \quad \text{and} \quad x = y^2 + 2y - 3$$

illustrate this and are given in Figure 18-2.

FIGURE 18-2

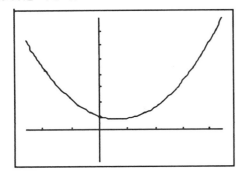 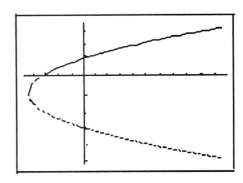

Recall also that if the axes of an ellipse are parallel to the two coordinate axes, then the ellipse has an equation of the form

$$\frac{(x - h)^2}{a^2} + \frac{(y - k)^2}{b^2} = 1$$

The center of the ellipse is at the point (h,k) and its vertices are the points

$$(h-a,k), \ (h+a,k), \ (h,k-b), \ (h,k+b)$$

This is illustrated by the graphs of

$$\frac{(x - 1)^2}{16} + \frac{(y - 2)^2}{9} = 1 \quad \text{and} \quad \frac{(x + 3)^2}{4} + \frac{(y - 1)^2}{25} = 1$$

given in Figure 18-3.

FIGURE 18-3

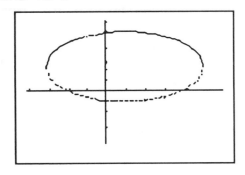 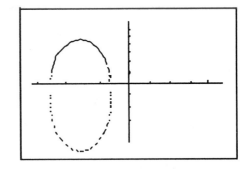

If the axis of a hyperbola is parallel to the x-axis, the curve has an equation of the form

$$\frac{(x-h)^2}{a^2} - \frac{(y-k)^2}{b^2} = 1$$

If the axis is parallel to the y-axis, the equation is of the form

$$\frac{(y-k)^2}{a^2} - \frac{(x-h)^2}{b^2} = 1$$

This is illustrated by the graphs of

$$\frac{(x-2)^2}{9} - \frac{(y+1)^2}{4} = 1 \quad \text{and} \quad \frac{(y+3)^2}{16} - \frac{(x+1)^2}{25} = 1$$

in Figure 18-4.

FIGURE 18-4

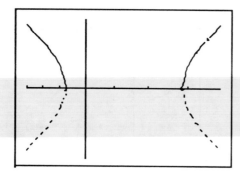 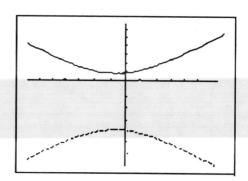

IN GENERAL:

The graph of an equation of the form

$$Ax^2 + By^2 + Cx + Dy + E = 0$$

is a conic section with axes parallel to the coordinate axes providing such a graph exists. If either A = 0 or B = 0 but not both, then it is a parabola. If AB > 0, it is an ellipse. If AB < 0, it is a hyperbola.

The basic technique for drawing graphs of conic sections and other complicated equations is to break the curve up into portions that are graphs of functions.

Example 1. Draw the graph of the parabola

$$y = .5x^2 - 3x + 1$$

SOLUTION: In this case, the equation represents a function and we only need to enter it and draw. The graph is given in Figure 18-5.

FIGURE 18-5

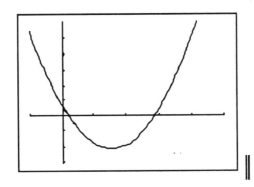

Graphs of conic sections that are not expressed as functions can be drawn by solving the equation for y in terms of x. Frequently this will involve completing the square.

Example 2. Draw the graph of

$$x = y^2 - 4y + 7$$

SOLUTION: We complete the square on y:

$$x - 3 = y^2 - 4y + 4$$
$$x - 3 = (y - 2)^2$$
$$y - 2 = \sqrt{x-3} \quad \text{or} \quad y - 2 = -\sqrt{x-3}$$
$$y = \sqrt{x-3} + 2 \quad \text{or} \quad y = -\sqrt{x-3} + 2$$

We now have two equations of functions whose graphs together comprise the graph of the original equation. Set Y_1 equal to the first, set Y_2 equal to the second, and draw. The result is given in **Figure 18-6.**

FIGURE 18-6

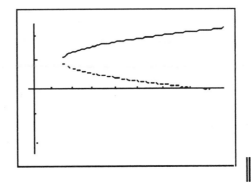

Example 3. Draw the graph of
$$4x^2 + 5x + y^2 - 8y = 34$$

SOLUTION: The graph is an ellipse because $(4)(1) > 0$.

First of all we rewrite the equation with just the y terms on one side.
$$y^2 - 8y = 34 - 5x - 4x^2$$

Now complete the square on y.
$$y^2 - 8y + 16 = 50 - 5x - 4x^2$$
$$(y - 4)^2 = 50 - 5x - 4x^2$$

Solve for y:
$$y - 4 = \sqrt{50-5x-4x^2} \quad \text{or} \quad y - 4 = -\sqrt{50-5x-4x^2}$$
$$y = \sqrt{50-5x-4x^2} + 4 \quad \text{or} \quad y = -\sqrt{50-5x-4x^2} + 4$$

Now draw the graphs of the two functions as in Figure 18-7.

FIGURE 18-7

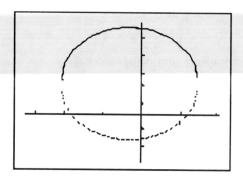

Some of the points on the graph were not pictured in the last example. The approximation process for square roots may cause your calculator to leave out a few points. ‖

Example 4. Draw the graph of
$$x^2 - 7x - 2y^2 - 12y = 6$$

SOLUTION: The graph is a hyperbola because $(1)(-2) < 0$.

Again we rewrite with just the y terms on one side:
$$2y^2 + 12y = x^2 - 7x - 6$$
$$2(y^2 + 6y) = x^2 - 7x - 6$$
$$(y^2 + 6y + 9) = .5x^2 - 3.5x - 3$$
$$(y + 3)^2 = .5x^2 - 3.5x - 3$$
$$y = \sqrt{.5x^2-3.5x-3} - 3 \quad \text{or} \quad y = -\sqrt{.5x^2-3.5x-3} - 3$$

The graphs of the two functions are given in Figure 18-8.

FIGURE 18-8

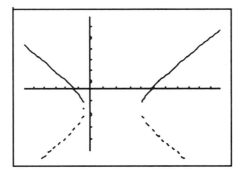

The vertices may not appear on the graph; again this is because we are dealing with approximations of square roots. ‖

Example 5. Draw the graph of
$$y^2 - 10y - x^2 + 6x = 5$$

SOLUTION: The graph is a hyperbola because $(1)(-1) < 0$.

First complete the square on y.
$$y^2 - 10y = 5 - 6x + x^2$$
$$y^2 - 10y + 25 = 30 - 6x + x^2$$
$$(y - 5)^2 = 30 - 6x + x^2$$
$$y = \sqrt{30-6x+x^2} + 5 \quad \text{or} \quad y = -\sqrt{30-6x+x^2} + 5$$

The graphs of both are given in Figure 18-9. The graph is drawn with a RANGE setting of $[-10,20,2,-20,20,2,1]$

FIGURE 18-9

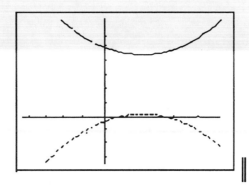

Hyperbolas provide examples of curves that have asymptotes that are neither vertical nor horizontal.

Example 6. The graph of
$$y^2 - x^2 = 1$$
is drawn by drawing the graphs of

$$y = \sqrt{1 + x^2} \quad \text{and} \quad y = -\sqrt{1 + x^2}$$

For values of x large in absolute value, the curve gets close to the lines
$$y = x \quad \text{and} \quad y = -x$$
as in Figure 18-10.

FIGURE 18-10

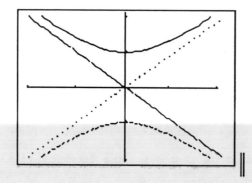

In general, the graphs of the hyperbolas

$$\frac{x^2}{a^2} - \frac{y^2}{b^2} = 1 \quad \text{and} \quad \frac{y^2}{b^2} - \frac{x^2}{a^2} = 1$$

have the lines

$$y = \frac{b}{a} x \quad \text{and} \quad y = -\frac{b}{a} x$$

as asymptotes.

Example 7. The lines

$$y = \frac{2}{3} x \quad \text{and} \quad y = -\frac{2}{3} x$$

are asymptotes of the graph of
$$\frac{x^2}{9} - \frac{y^2}{4} = 1.$$

The graphs of the lines and the curve are given in Figure 18-11.

FIGURE 18-11

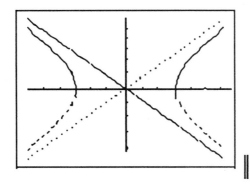

The circumference of an ellipse can be approximated using numerical integration and the formula given in Section 15.

Example 8. Compute the circumference of the ellipse

$$\frac{x^2}{4} + \frac{y^2}{16} = 1$$

SOLUTION: First we solve for y in terms of x and draw the graph in Figure 18-12.

$$y = \pm 2(4 - x^2)^{1/2}$$

FIGURE 18-12

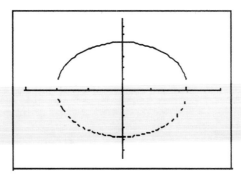

We will approximate the length of $y = 2(4 - x^2)^{1/2}$ and double it to get the circumference.

The derivative is

$$y' = -2x(4 - x^2)^{-1/2}$$

The formula for arc length tells us that

$$L = \int_{-2}^{2} \sqrt{1 + (y')^2} \, dx$$

Substituting for y' and doing some algebra yields

$$L = \int_{-2}^{2} \sqrt{1 + \frac{4x^2}{4 - x^2}} \, dx$$

Since the denominator does not exist for $x = \pm 2$, we must run our approximation from numbers close to -2 and 2. We use -1.99 and 1.99 and approximate the integral by the Trapezoidal Rule with 64 subdivisions. The results are displayed on the screen in Figure $18-13$. We have multiplied the approximation by 2 to approximate the entire circumference of approximately 18.56 units.

FIGURE 18-13

```
?-1.99
ENTER B
?1.99
ENTER N
?64
APPROX IS
      9.279729815
Ans*2
      18.55945963
```

Other calculus problems involving conic sections can be solved using techniques developed earlier. We include some of these in the exercises.

EXERCISES

1. Draw the graph of
$$y = .3x^2 - 2x + 1$$

2. Draw the graph of
$$x = 2y^2 - 12y + 3$$

3. Draw the graph of
$$x^2 + y^2 = 19$$

4. Draw the graph of

$$x^2 - y^2 = 10$$

5. Draw the graph of

$$y^2 - x^2 = 2$$

6. Draw the graph of

$$x^2 + 3x + y^2 = 12$$

7. Draw the graph of

$$x^2 - 7x - 2y^2 = 3$$

8. Draw the graph of

$$3x^2 - x + y^2 - 6y = 20$$

9. Draw the graph of

$$5x^2 + 2x - y^2 + 5 = 16$$

10. Draw the graph of

$$y^2 + 4y - 2x^2 - 7x = 11$$

11. Draw the graphs of

$$y = .5x^2 \qquad and \qquad y = .5x^2 - 3x - 1$$

What can you say about the two curves?

12. Draw the graphs of

$$4x^2 + 9y^2 = 36 \qquad and \qquad 4x^2 + 9y^2 = 144$$

on the same coordinate system. What can you say about the curves?

13. Draw the graphs of

$$2x^2 + 5y^2 = 10 \qquad and \qquad 5x^2 + 2y^2 = 10$$

on the same coordinate system. What can you say about the curves?

14. Draw the graphs of

$$3x^2 + 11y^2 = 48 \qquad and \qquad 3x^2 - 6x + 11y^2 - 44y = 1$$

on the same coordinate system.   What can you say about the curves?

15. Use numerical differentiation to approximate the slope of the line tangent to the graph of

$$\frac{x^2}{10} + \frac{y^2}{15} = 1$$

at the point $(2,3)$.

16. Use numerical differentiation to approximate the slope of the line tangent to the graph of

$$\frac{x^2}{16} - \frac{y^2}{25} = 1$$

at the point where $x = 10$ and y is positive.   How does this number compare to the slope of the asymptote?

17. Use numerical integration to approximate the length of the parabola

$$y = 2x^2 - 3x + 7$$

from $x = 2$ to $x = 5$.

18. Use numerical integration to approximate the length of the upper part of the hyperbola

$$\frac{y^2}{16} - \frac{x^2}{9} = 1$$

from $x = -3$ to $x = 3$.

19. Use the ZERO program or an equivalent process to approximate the points of intersection of

$$\frac{(x-1)^2}{5} + \frac{(y+3)^2}{14} = 1 \quad \text{and} \quad \frac{x^2}{2} - \frac{(y+2)^2}{3} = 1$$

20. Draw graphs of four ellipses with center at the origin and use a Square RANGE setting.   Compare the denominators and the "roundness" of the ellipses.   Make a conjecture saying something about the "roundness" of an ellipse.

19

PARAMETRIC EQUATIONS AND POLAR COORDINATES

It is not necessary to have y expressed in terms of x in order to draw the graph of a curve. Your calculator can be used to draw graphs of curves if both x and y are expressed in terms of a third variable called a *parameter*. (In fact it will draw the graph of more than one parametric curve on the coordinate system.) You need to express both x and y in terms of the parameter.

Example 1. Draw the graph of
$$x = t^2 + t + 1 \qquad \text{and} \qquad y = t^3 - t^2 + 2t + 2$$

SOLUTION: First of all, your calculator needs to be set in the parametric mode. For the TI-81 this is done by pressing $\boxed{\text{MODE}}$ and then selecting $\textsf{Param}$. In the examples we use the notation of the TI-81.

Let
$$X_{1T} = T^2 + T + 1$$

and
$$Y_{1T} = T^3 - T^2 + 2T + 1$$

The $\textsf{RANGE}$ setting now involves three variables T, X_{1T}, and Y_{1T}.

You need to set the limits and the size of steps for the parameter as well as for X_{1T} and Y_{1T}. For this example, let T run from -5 to 5 with steps of .1. From the nature of the function, we know that X_{1T} will not take any negative values but to get a better view we let the minimum value of X_{1T} be 0. For $T = 5$, $X_{1T} = 31$ and 35 seems a reasonable maximum with a choice of 5 for the X_{1T}-scale. We can expect Y_{1T} to run from about -170 to 170 and a scale of 20 seems reasonable. The RANGE setting is [-5,5,.1,-5,35,5,-170,170,20]; the graph is given in Figure 19-1. Note that the curve is not the graph of a function because one value of x corresponds to two values of y in many cases.

FIGURE 19-1

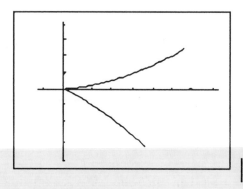

When using parameters, TRACE starts at T = 0 and shifting the cursor is done by decreasing or increasing the values of T. The values of all three variables are displayed. We could use TRACE to estimate the value of T where the curve crosses the x-axis but applying the ZERO program or a similar process to

$$y = t^3 - t^2 + 2t + 1$$

will give better accuracy.

Example 2. Draw the graph of
$$x = \sin(t^2-2)$$
$$y = \cos(t^3+2t-1)$$

SOLUTION: In this case, the x and y values both vary between -1 and 1 and the periodicity of the trigonometric functions leads us to expect the same x and y values to appear more than once. Let us use a RANGE setting of [-2,2,.1,-1.2,1.2,1,-1.2,1.2,1].

Then $\qquad\qquad X_{1T} = \sin(T^2-2)$

and $\qquad\qquad Y_{1T} = \cos(T^3+2T-1)$

The graph is given in Figure 19-2.

FIGURE 19-2

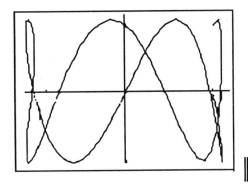

The graph of any function can also be described using parameters. If we let

$$x = t \qquad \text{and} \qquad y = f(t)$$

then we get the same graph as we get for
$$y = f(x).$$

Some quite complicated curves can be expressed rather simply using parameters.

Example 3. A cycloid is the path taken by a point on the outside of a wheel as the wheel rolls along. If the point starts at the origin, the radius of the wheel is 10, and t is the angle that the point has turned through, the coordinates of the point are

$$X_{1T} = 10T - 10\sin T$$

$$Y_{1T} = 10 - 10\cos T$$

The graph is given in Figure 19-3.

FIGURE 19-3

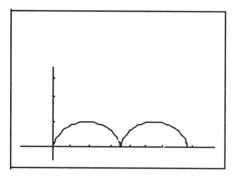

The general equations for a cycloid are obtained from those above by replacing 10 by R, the radius of the wheel.

$$X_{1T} = RT - R\sin T$$

$$Y_{1T} = R - R\cos T$$

The path of an object can often be described by parametric equations that give its position variables in terms of one variable t.

Example 4. Suppose an object is dropped from a moving airplane at a height of 2500 feet. Its coordinates at time T might be given by

$$X_{1T} = 450 - 5T - .2T^2$$

$$Y_{1T} = 2500 - 16T^2$$

The graph is given in Figure 19-4.

FIGURE 19-4

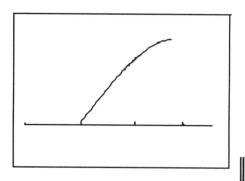

Slopes of tangent lines to curves given parametrically can be approximated by using the numerical differentiation process twice and then dividing. You need to use the Chain Rule in the form:

$$\frac{dy}{dx} = \frac{dy/dt}{dx/dt}$$

Example 5. Approximate the slope of the curve

$$X_{1T} = \ln[T^2 + e^{-2T}] + 5T$$

$$Y_{1T} = \cos[3T - 7 + \ln(T^2+1)]$$

at the point $T = 1.7$.

SOLUTION: The graph is given in Figure 19-5.

FIGURE 19-5

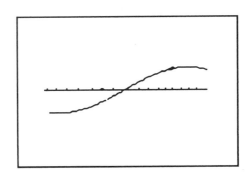

First we approximate $\dfrac{dX_{1T}}{dT}$ and $\dfrac{dY_{1T}}{dT}$ at $T = 1.7$. (The calculator must be in function mode to use the numerical differentiation process.) We then divide to approximate the derivative. The results are displayed in the screens in Figure 19-6.

FIGURE 19-6

```
1.7→K
                1.7
NDeriv(ln|x²+e^
(-2X00+5X,.01)
        6.140214239
Ans→H
        6.140214239
```

```
NDeriv(cos(3T-7
+ln(K²+1)),.01)
        1.996606696
Ans→K
        1.996606696
H/K
        3.075324876
```

The slope of the tangent line is approximately 3.0753. ▌

Most graphics calculators use parameters to draw graphs of functions in polar coordinates. Recall that the x and y coordinates of a point with polar coordinates [r,θ] are

$$x = (\cos θ)r$$
$$y = (\sin θ)r.$$

If we use the variable T instead of θ, the graph of any function

$$r = f(θ)$$

can be drawn using the parametric equations

$$X_{1T} = (\cos T)f(T)$$

$$Y_{1T} = (\sin T)f(T)$$

Remember that it is necessary to determine a RANGE setting involving all three variables.

Example 6. One of the simplest graphs to draw in polar coordinates is the spiral

$$r = θ$$

The parametric equations are

$$X_{1T} = (\cos T)T$$

$$Y_{1T} = (\sin T)T$$

If we let T run from 0 to 4π or about 12.6, then x and y will run from about -12.6 to 12.6. We use the RANGE setting of [0,12.6,.1,-18.9,18.9,4,-12.6,12.6,2] to make the system square. The graph is given in Figure 19-7.

FIGURE 19-7

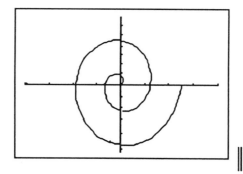

Example 7. Draw the graph of the three leaf rose

$$r = 2\sin(3\theta)$$

SOLUTION: In this case, we know that the function is periodic and will repeat after T runs from 0 to 2π. Since neither x nor y can be greater than 2, a RANGE setting of $[0,6.29,.1,-3.75,3.75,1,-2.5,2.5,1]$ will suffice and provide a square system.

Let $\qquad\qquad X_{1T} = (\cos\ T)(2\sin(3T))$

$\qquad\qquad\qquad\quad Y_{1T} = (\sin\ T)(2\sin(3T))$

The graph is given in Figure 19-8

FIGURE 19-8

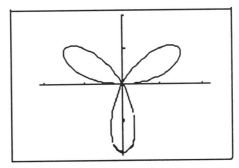

You may find it interesting to change the range so that T runs from 0 to π. ▌

Conic sections other than circles can be described using a point F called the *focus* and a line L called the *directrix*. The idea is illustrated in Figure 19-9.

FIGURE 19-9

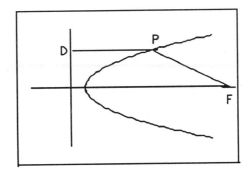

The *eccentricity* e of a conic section is defined by the constant ratio

$$e = \frac{PF}{PD}$$

where PF is the distance from a point P on the curve to the focus and PD is the distance from the same point to the directrix. It can be shown that, for such conic sections, the equation in polar coordinates with the focus at the origin and directrix parallel to the y-axis is

$$r = \frac{ke}{1 + e\cos\theta}$$

where e is the eccentricity of the curve.

Example 8. Draw the graph of

$$r = \frac{6}{1 + 2\cos\theta}$$

SOLUTION: The parametric equations are

$$X_{1T} = (\cos T)\left[\frac{6}{1 + 2\cos T}\right]$$

$$Y_{1T} = (\sin T)\left[\frac{6}{1 + 2\cos T}\right]$$

with a RANGE setting of $[0, 2\pi, .1, -12, 12, 1, -12, 12, 1]$ The graph is given in Figure 19-10.   Note that the eccentricity is 2; the graph is a hyperbola.   (The TI-81 draws the asymptotes automatically.)

FIGURE 19-10

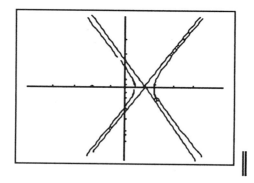

In analytic geometry it is shown that:

    if $0 < e < 1$, the curve is an ellipse

    if $e = 1$. the curve is a parabola

    if $1 < e$, the curve is a hyperbola

The closer e is to zero, the rounder the ellipse, i.e. the nearer the curve is to being a circle.

Example 9. Draw the graph of

$$r = \frac{3}{1 + .5\cos \theta}$$

SOLUTION: The parametric equations are

$$X_{1T} = (\cos T)\left[\frac{3}{1 + .5\cos T}\right]$$

$$Y_{1T} = (\sin T)\left[\frac{3}{1 + .5\cos T}\right]$$

The graph is given in Figure 19-11 with the Standard RANGE setting.

FIGURE 19-11

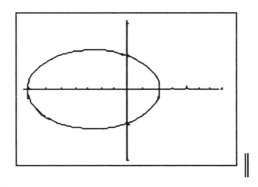

The graph of an equation of the form

$$r = a + b\cos \theta$$

is called a *limicon* if a ≠ b. It has a loop if $0 < a < b$ and a "dent" if $0 < b < a$. The curve is called a *cardiod* if $0 < a = b$. The three types are illustrated in Figure 19-12.

FIGURE 19-12

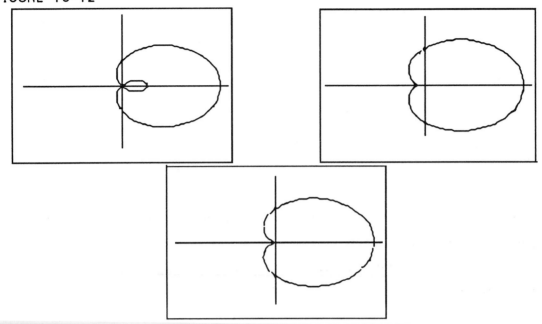

In the second graph, the curve does not include the origin but in the third it does. This is because $x = y = 0$ for $t = \pi$ if $b = a$, but for $b < a$, $x < 0$ when $y = 0$ and $t = \pi$.

The "roses" are special curves with rather simple equations in polar coordinates. The graph in Example 7 was a three leaf rose. The next example has four leaves.

Example 10. Draw the graph of
$$r = 3\cos(2\theta)$$

SOLUTION: The parametric equations are

$$X_{1T} = (\cos T)(3\cos 2T)$$

$$Y_{1T} = (\sin T)(3\cos 2T)$$

We use $[0, 6.28, .1, -4.5, 4.5, 1, -3, 3, 1]$ for the RANGE setting. The graph is given in Figure 19-13.

FIGURE 19-13

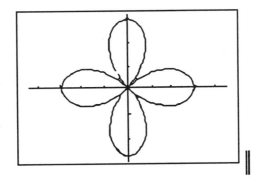

The techniques and formulas used with functions of one variable have analogues for parametric curves.

The formula for computing arc length for a curve given parametrically is

$$L = \int_a^b \sqrt{(dx/dt)^2 + (dy/dt)^2}\ dt$$

Example 11. Approximate the length of the curve

$$x = t^2 + t + 4$$
$$y = 2t^3 + \sin(\pi t)$$

from $t = 0$ to $t = 4$.

SOLUTION: The graph is given in Figure 19-14.

FIGURE 19-14

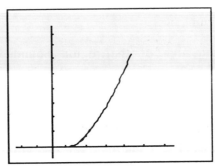

First compute the two derivatives:

$$\frac{dx}{dt} = 2t + 1$$

$$\frac{dy}{dt} = 6t^2 + \pi\cos(\pi t)$$

Since the programs for numerical integration are written in functional notation, we let Y_1 be the integrand and change the variable to X.

$$L = \int_0^4 \sqrt{(2X+1)^2 + (6X^2 + \pi\cos(\pi X))^2} \, dX$$

The results of using Simpson's Rule with 128 subdivisions are given in Figure 19-15.

FIGURE 19-15

```
?0
UPPER LIMIT
?4
N DIVISIONS
?128
INTEGRAL-
        130.5317838
```

The length is about 130.53 units. ‖

The area of the region enclosed by the two rays $\theta = a$ and $\theta = b$ with $0 \le a \le b \le 2\pi$ and the graph of the non-negative curve $r = f(\theta)$ is given by

$$A = \frac{1}{2} \int_a^b [r(\theta)]^2 \, d\theta$$

Example 12. Approximate the area of the region bounded by the graph of
$$r = \theta^2 + 2\sin(\cos(.3\theta)) + 2$$
and the rays $\theta = .4$ and $\theta = 1.2$.

SOLUTION: The parametric equations for the curve are

$$X_{1T} = (\cos T)(T^2 + 2\sin(\cos(.3T)) + 2)$$

$$Y_{1T} = (\sin T)(T^2 + 2\sin(\cos((.3T)) + 2)$$

The lines are given by

$$X_{2T} = (\cos .4)T$$

$$Y_{2T} = (\sin .4)T$$

and

$$X_{3T} = (\cos 1.2)T$$

$$Y_{3T} = (\sin 1.2)T$$

The region is drawn in Figure 19-16.

FIGURE 19-16

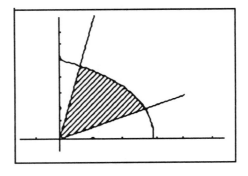

The area is given by

$$A = \frac{1}{2} \int_{.4}^{1.2} [X^2 + 2\sin(\cos(.3X)) + 2]^2 \quad dx$$

We use the Trapezoidal Rule with 64 subdivisions. The results are displayed on the screen in Figure 19-17.

FIGURE 19-17

```
?1.2
ENTER N
?64
APPROX IB
        15.1810031
Ans/2
        7.59050154B
```

Notice that we divided the approximation by 2 since the formula indicates that we are to multiply the integral by 1/2. The area is approximately 7.59 square units. ‖

Other applications of calculus can be used with parametric or polar equations. In most cases, the formulas need to be adjusted as were the formulas for arc length and area.

EXERCISES

1. Draw the graph of $x = \sin t$ and $y = \cos t$.

2. Draw the graph of $x = t^2$ and $y = t^3 + 1$

3. Draw the graph of $x = e^{-t}$ and $y = e^t$. What is the shape? Why?

4. Draw the graph of $x = 3t + 2$ and $y = 2t - 1$. What is the shape? Why?

5. Draw the graph of $x = e^{2t} - 1$ and $y = e^{-t} + 2$

6. Draw the graph of $x = t$ and $y = \cosh t$. Does it look familiar? What does it look like? Why?

7. Draw the graph of $x = 5\sin^{-1} t$ and $y = 5\cos^{-1} t$. What is the shape? Why?

8. Draw the graph of $x = \tan^{-1} t$ and $y = \tan t$

9. Choose four non-zero numbers a, b, c, and d. Draw the graph of
$$x = at + b \text{ and } y = ct + d$$
What can you say about the graph? Can you express y in terms of x?

10. Draw the graph of $x = \cos t$ and $y = \sec t$. Is the shape of the curve familiar? Draw the graph of $x = \tan t$ and $y = \cot t$. What can you say about the two graphs? Express y in terms of x.

11. Draw the graph of $r = 4\sin \theta$

12. Draw the graph of $y = 3\cos \theta$

13. Choose two numbers a and b and draw the graphs of the two functions
$$r = a\sin\theta \quad \text{and} \quad r = b\cos\theta$$
on the same coordinate system using a Square RANGE setting.
What can you say about the graphs?

14. Choose two numbers a and b and draw the graphs of
$$r = a\sec\theta \quad \text{and} \quad r = b\csc\theta$$
What can you say about the graphs?

15. Choose a number a and draw the graph of $r = a$. What is its shape?
(First ZOOM to Square.) What is the shape? Explain why.

16. Draw the graphs of the following:

    (a) $r = \dfrac{2}{1 + .4\cos\theta}$ (b) $r = \dfrac{3}{1 + \cos\theta}$

17. Draw the graphs of the following:

    (a) $r = \dfrac{4}{1 + 3\cos\theta}$ (b) $r = \dfrac{8}{3 + 2\cos\theta}$

18. Draw the graphs of the following:

    (a) $r = \dfrac{2}{1 + 3\sin\theta}$ (b) $r = \dfrac{5}{1 - \sin\theta}$

19. Let $x = e^{-2t} + t^2$ and $y = \cosh(t^3 + t + 1) + t^3$
    Approximate $\dfrac{dy}{dx}$ for $t = 2$.

20. Let $x = \dfrac{\sin(3t) - \ln(t^2 + 5)}{2 + \cos t}$ and $y = \dfrac{t^2 + e^{\cdot 4t}}{3 + \sin t}$.
    Approximate $\dfrac{dy}{dx}$ for $t = 2$.

21. Draw the path of an object that is located at $400t$ meters east
    and $100t - 16t^2$ meters north of the starting point at time t.

22. Draw the graph of the path of an object that is located $e^t + 5t - 1$ miles east of the starting point and $4\sin t + 5t$ miles north of the starting point at time t.

23. Draw the path of a moving object that has an eastward velocity of $[4t + \sin t]$ mph and a northward velocity of $[3t^2 + \cosh t]$ mph at time t and is at the origin at time $t = 0$.

24. Choose a function $f(t)$. Suppose a curve has parameters such that
$$\frac{dx}{dt} = f(t) \quad \text{and} \quad \frac{dy}{dt} = f(t).$$
Draw the graph of the curve. Choose another function $f(t)$ and draw the graph. State a general conclusion and justify your answer.

25. Draw the graphs of $r = 3\sin(4\theta)$ and $r = 2\sin(5\theta)$.

25. Choose two numbers m and n, one even and one odd. Draw the graphs of
$$r = \cos(m\theta) \text{ and } r = \cos(n\theta)$$
What can you conclude about the graphs?

27. Draw the graphs of the following:
    (a) $r = 4 + 3\cos\theta$ (b) $r = 3 + 4\cos\theta$

28. Draw the graphs of the following:
    (a) $r = 3 + 3\cos\theta$ (b) $r = 4 + 2\sin\theta$

29. Draw the graph of the following for $0 \leq \theta \leq 4\pi$:
    (a) $r = \cos\dfrac{\theta}{2}$ (b) $r = \sin\dfrac{\theta}{2}$

30. Draw the graphs of the following:
    (a) $r = \cos^3(2\theta)$ (b) $r = \sin^3(2\theta)$

27. Choose a function r of θ that involves no non-constant function other than the cosine. Draw its graph. Now substitute sine for every occurance of cosine and draw the graph of the new function on the same coordinate system. How are the graphs alike? Why?

20

CURVE FITTING

In many problems in science and commerce, data arises from measurements or experiments. In most cases, we get a better grasp of the situation if we can describe it using some elementary function. One way of doing so is to attempt to fit the data points with the graph of a function as in Figure 20-1.

FIGURE 20-1

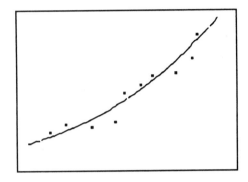

We can always just draw the line segments connecting the points but this is seldom the graph of an elementary function. Such lines have "corners" on them and therefore do not have derivatives at these points. We would like to approximate the behavior by a smooth curve that is the graph of a relatively simple function.

The TI-81 calculator uses portions of its statistics package to approximate collections of data by four types of functions:

linear	$Y = a + bX$
logarithmic	$Y = a + b\ln X$
exponential	$Y = ab^X$
power	$Y = aX^b$

The calculator will determine a and b for each type of function and also compute a number r, the regression coefficient, that gives some measure of how well the function fits the data.

The closer the absolute value of r is to 1, the better the fit.

Example 1. Suppose that the population data for a city is given in the following table.

TABLE 1	year	population
	1960	162,400
	1965	167,500
	1970	172,300
	1975	178,000
	1980	184,600
	1985	192,500
	1990	199,800

Approximate the population function by a function of time in one of the above forms.

SOLUTION: First of all, let us measure time starting in 1955. Let 1955 correspond to 0, 1960 correspond to 5 and so on. The x-values will be the dates and the y-values will be the population. On the TI-81, we use 2nd , STAT , DATA, 1, and then enter the x-values and the y-values. The entries are displayed on the screens in Figure 20-2.

FIGURE 20-2

```
DATA
X1=5
Y1=162400
X2=10
Y2=167500
X3=15
Y3↓172300
```

```
X4=20
Y4=178000
X5=25
Y5=184600
X6=30
Y6=192500
X7=35
Y7↓199800
```

Once these have been entered, we can use DRAW and 1, 2, or 3 to obtain a histogram, a scatter diagram, or a line connecting the sequential points. All three are displayed in Figure 20-3. (If the data points have not been entered in increasing order of x-values, you should use xSort first before drawing the graphs.) For this data, the RANGE setting used will be [0,40,5,160000,200000,10000,1]; we set the maximum and minimum values for each variable to include the largest and smallest values in the table and set the scales accordingly.

FIGURE 20-3

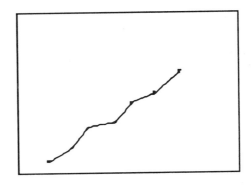

We now have four approximating functions determined by the calculator; they are described by the screens in Figure 20-4.

FIGURE 20-4

```
2
LinReg
 a-154557,142B
 b-1246,428571
 r-.9B545825B
```

```
3
LnReq
 a-127371,9072
 b-18467,50964
 r-.93508D701 5
```

```
4
Exp Reg
 a-155994,1092
 b-1.006945028
 r-.9B7726723
```

```
5
PwrReg
 a-133759,8685
 b-.1033461784
 r-.9445763117
```

From the values of r, we can see that the exponential function

$$Y = 155994.1092(1.006935028)^X$$

is a good fit and the linear function is almost as good. The other two do not fit the data quite as well. The graph of this function and the scatter diagram given in Figure 20-5 illustrate how well the function fits the data.

FIGURE 20-5

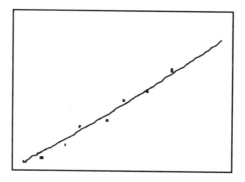

One use of such functions is to predict other values. We can use this function to predict the population in 1995.

For   X = 40,   Y ≈ 205,668

and we can expect the population to be about 205,688. ‖

In an experiment, an elementary function seldom is an exact fit to the data   In most cases we are just looking for an approximation.

Example 2. Suppose that the number of students and number of faculty at a university for a five year period is given by the following table.

TABLE 2	students	faculty
	12,352	617
	12,568	625
	13,110	630
	13,340	633
	13,800	635

Which of the four types of functions available with the TI-81 is the best fit?

SOLUTION: Testing the four types of functions, we see from Figure 20-6 that the best fit is the logarithmic function
$$Y = -818.1264039 + (152.6333417)\ln X$$
with regression coefficient r = 0.949200528

Actually all four seem to be rather good.

FIGURE 20-6

```
2
LinReg
  a=476.1388434
  b=.0116511552
  r=.9450427802
```

```
3
LnReg
  a=-818.126403B
  b=152.6333417
  r=.949200528
```

```
4
ExpReg
  a=492.8251231
  b=1.000018592
  r=.9437872005
```

```
5
PwrReg
  a=62.47191165B
  b=.243576671B
  r=.9479927632
```

Both the function and the scatter diagram are shown in Figure 20-7.

FIGURE 20-7

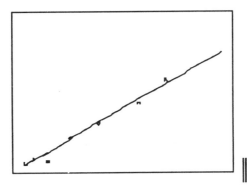

Example 3. The following data was obtained from the U.S. Bureau of Labor Statistics.

TABLE 3	year	potential labor force (in thousands)
	1970	140182
	1971	145596
	1972	145775
	1973	148263
	1974	150827
	1975	153449
	1976	156048
	1977	158559
	1978	161058
	1979	163620

Which of the four available functions provides the best fit?

SOLUTION: The screens in Figure 20-8 show that the linear function

$$Y = -32364.95785 + 2479.230303X \qquad \text{for} \quad 70 \leq x \leq 79$$

is the best fit with a regression coefficient of $r = 0.9949956071$

although all of them are quite good.

FIGURE 20-8

2	3
LinReg	LnReg
a=-32364.9575B	a=-642725.7215
b=2479.230303	b=184467.1202
r=.9949966071	r=.9947474052

4	5
ExpReg	PwrReg
a=45189.23002	a=B16.0677863
b=1.0164062B	b=1.213035844
r=.9943204855	r=.9944400772

If we draw the graph of this function and the scatter drawing with
RANGE setting [70,80,1,145000,175000,10000,1], all but one of the

data points seem to be on the graph.

FIGURE 20-9

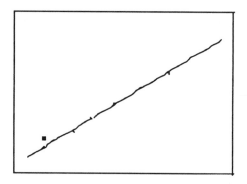

If we try to use this function to predict the potential labor force in
1980, we get $Y \approx 165973$.   The data source used to obtain the table

gives the actual force at 166789.   All this means is that the change in the labor force for 1980 was not quite what might have been expected using the data from previous years. There may have been some special occurance that changed the pattern. (The entry of the post WWII baby boom into the work place might have made the difference.) ‖

Example 4. An appliance store found that the number of TV sets sold at several prices was given by the following table:

TABLE 4

Price	number of sets sold
200	825
250	755
300	611
350	503
400	361
450	321
500	267
550	255
600	218
650	172
700	104
750	78
800	51
900	32
950	11

Which of the four available functions gives the best approximation of the function that relates prices and sales?

SOLUTION: The scatter diagram is given in Figure 20-10.

FIGURE 20-10

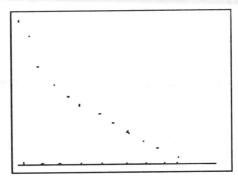

By looking at the four functions and the regression coeficients in Figure 20-11, we see that the best approximation is given by the logarithmic function.

FIGURE 20-11

```
2
LinReg
 a=891.065208
 b=1.054129116
 r=.9457620851
```

```
3
LnReg
 a=3707.547982
 b=546.5932818
 r=-.9898212834
```

```
4
ExpReg
 a=3155.71552
 b=.9948922041
 r=-.970012207
```

```
5
PwrReg
 a=505780893
 b=-2.382658975
 r=-.9109581897
```

The graph of the logarithmic function

$$Y = 3709.91765 + (-546.9420962)\ln X$$

along with the scatter diagram in Figure 20-12 shows just how good a fit we have.

FIGURE 20-12

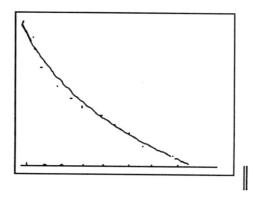

Sometimes data cannot be approximated well by one of the functions available with the TI-81 calculator.

Example 5. Suppose our data consists of the following points:

TABLE 5	X	Y
	1	.8
	2	.9
	3	.1
	4	-.8
	5	- 1

If we enter the data and try the four approximating functions, we see that only the linear and logarithmic approximations are possible since Y takes both positive and negative values.

The linear function

$$Y = 1.615 - .527X$$

is the best fit with a regression coefficient of $r = -0.9547602807$ but if we draw the graph and the scatter diagram as in the first drawing of Figure 20-13 we see that it really is not a good fit. In the second drawing we can see that the sine curve fits this data quite well but it is not within the capacity of the instrument to determine such a function. There are programs available for use with computers that would give quite a good fit.

FIGURE 20-13

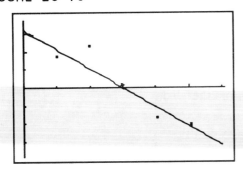 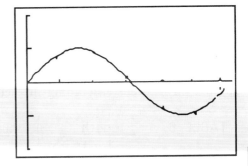

EXERCISES:

In Exercises 1-6, enter the data, study the scatter drawing, and see which of the functions given by the calculator is the best fit. The data points are given as sets of ordered pairs instead of in a table. You will have to determine each RANGE setting. Draw the graphs of the functions and the scatter diagrams.

1. (1,5), (2,20), (3,41),(4,79), (5, 330)

2. (10,19), (12,23), (14,28),(16,35), (18,40)

3. (5,12), (7,13), (10,19), (15,27)

4. (2.3,1.76),(2.7,1.93), (3.1,2.7), (3.8,4.1), (4.8,5.1)

5. (-1.5,6.7), (4.3,1.6), (2.3,3.7), (3.4,2.5)
    You will get an error message when you try 3. Why is no curve of this type available?

6. (-6,0),(-4,25),(-2,35),(1,50),(4,73),(6,87)   Why are some of the curves not available for this example?

7. The following table gives the number of bass caught by a group of five anglers in a week according to length of the fish.

TABLE 6	length	number caught
	12	34
	13	31
	14	27
	15	24
	16	18
	17	15
	18	10
	19	7
	20	4
	21	1
	22	1

Determine the best fit of the types of curves available.

8. The following table gives the annual student fees for a period of ten years at a state university. Find a function of the forms discussed that gives the best fit for this data. Draw the graphs of the function and the scatter diagram. Use the function to predict the fees five years from the last year given.

TABLE 7	year	fees
	1976	1873
	1977	2010
	1978	2245
	1979	2395
	1980	2615
	1981	2920
	1982	3155
	1983	3410
	1984	3700
	1985	3950

9. The following table gives populations of bacteria in a culture at several times in one day. Find a function of the forms available that is the best fit for this data. Use the function to predict the population at noon the next day.

TABLE 7

time	population
12:00 Noon	15000
02:30 PM	15,600
04:00 PM	15,900
08:00 PM	17,100
10:00 PM	18,200

10. Use an almanac or other source to determine the population of the earth each of the last ten years. Use this information to approximate a population function for the earth and predict the population five years from now.

11. The following table gives the standings and percentage of wins for the American League East and for the American League West on a day in July, 1991. Determine a curve of the type available that is the best fit for the data for eash division. Are the functions much the same? What can you infer from this information?

TABLE 8

EAST		WEST	
Standing	Percent wins	Standing	Percent wins
1	.591	1	.591
2	.516	2	.534
3	.500	3	.533
4	.495	4	.527
5	.451	5	.516
6	.407	6	.516
7	.344	7	.473

12. The following table contains the finishing positions and prize money for the first ten cars in the Toronto Indy in 1991. Determine a curve of the type available that is the best fit for the data.

TABLE 9

Position	Winnings
1	133,000
2	106,500
3	90,000
4	67,000
5	61,000
6	54,000
7	47,500
8	41,000
9	34,500
10	33,000

13. Use an almanac or other source to find the Cost of Living Index for the years 1981-1990. Make a table and determine a curve of the type available that is the best fit. Use it to predict the Cost of Living Index for 1993.

14. In the following experiment, you are to start with ten coins. The first column gives the number of coins in a collection and the second number is the number of heads obtained by tossing each collection of coins eight times and entering the total number of heads obtained in the eight tosses.

TABLE 10

number of coins	number of heads obtained
3	_____
4	_____
5	_____
6	_____
7	_____
8	_____
9	_____
10	_____

For example, if you toss a collection of 3 coins 8 times and obtain a total number of 13 heads, enter 13 opposite the 3 in the table.

Complete the table and find a function in one of the forms available that best fits the data. Draw the graph of the function and the scatter diagram. Use your function to predict the number of heads obtained by tossing 11 coins 8 times. Try it and see how close your prediction is to the number obtained.

15. Use Table 1 from Example 1 to determine a best fitting curve if you let time start at 1900, i.e. let t = 0 in 1900. Compare your functions to the functions obtained in the example. Especially compare the regression coefficients. How are the functions related? Explain your answer.

16. Add 10 to each x-value and 10 to each y value in the data from Exercise 6. Are all four curves available now? How do they compare to the ones you obtained previously? How are the new functions related to the original ones? Explain your answer.

APPENDIX

A-1: PROGRAMS FOR THE TI-81

1. ZERO: THE BISECTION PROCESS

```
:Pgm :ZERO                              :If abs(A-B)<E
:Disp"CHOOSE A,B SO F(A)F(B)<0"         :Goto 9
:Lbl 1                                  :If P*Y₁>0
:Disp"ENTER A"                          :Goto 7
:Input A                                :M→B
:Disp"ENTER B"                          :Goto 2
:Input B                                :Lbl 7
:.000000001→E                           :M→A
:A→X                                    :Goto 2
:Y₁→P                                   :Lbl 9
:B→X                                    :Disp"ZERO AT"
:If P*Y₁>0                              :Disp M
:Goto 8                                 :Disp "Y₁="
:Lbl 2                                  :Disp Y₁
:(A+B)/2→M                              :Goto 5
:A→X                                    :Lbl 8
:Y₁→P                                   :Disp"F(A)F(B)>0, CHOOSE AGAIN"
:M→X                                    :Goto 1
                                        :Lbl 5
                                        :End
```

2.   LIMIT OF A FUNCTION

```
Prgm :LIMIT
:10000→P
:-10000→Q
:0→N
:.000000001→E
:Disp"ENTER C"
:Input C
:.1→D
:Lbl 1
:C+D→X
:If abs(Y₁-P)<E
:Goto 6
:Y₁→P
:C-D→X
:If abs(Y₁-Q)<E
:Goto 7
:Y₁→Q
:D*.1→D
:1+N→N
:If N>12
:Goto 5
:Goto 1
:Lbl 5
:Disp "AFTER 12
APPROXIMATIONS N
O LIMIT"
:Goto 9
```

```
:Lbl 6
:Disp"RIGHT LIM
IT"
:Disp P
:Lbl 7
:Disp "LEFT LIM
IT"
:Disp Q
:If abs(P-Q)>10
0*E
:Goto 4
:(P+Q)/2→W
:Disp"LIMIT IS"
:Disp W
:Goto 9
:Lbl 4
:Disp"RIGHT AND
LEFT LIMITS DI
FFER OR ONE DOES
NOT EXIST, NO L
IMIT"
:Lbl 9
:End
```

3.   NEWTON'S METHOD

```
Prgm :NEWTON
:Disp "ENTER E"
:Input E
:Disp "ENTER X"
:Input X
:Lbl 1
:X-(Y₁/Y₂)→A
:If abs (X-A)<E
:Goto 2
:A→X
:Goto 1
:Lbl 2
:Disp "SOLUTION
IS"
:Disp A
:End
```

4.   TRAPEZOIDAL RULE:

```
Prgm  :TRAP                   :A→X
:Disp"ENTER A,B,              :1→M
AND N, NUMBER OF D            :Lbl 2
IVISIONS"                     :X+D→X
:Disp"ENTER A"                :Y₁+K→K
:Input A                      :1+M→M
:Disp "ENTER B"              :If M<N
:Input B                      :Goto 2
:Disp"ENTER N"                :K*D→K
:Input N                      :Disp"APPROX
:(B-A)/N→D                    IS"
:A→X                          :Disp K
:Y₁/2→K                       :End
:B→X
:Y₁/2+K→K
```

5. SIMPSON'S RULE:

```
Prgm :SIMPSON
:All-Off
:Disp"LOWER LIM
IT"
:Input A
:Disp"UPPER LIM
IT"
:Input B
:Disp"N DIVISIO
NS"
:Input D
:0→S
:(B-A)/2D→W
:1→J
:Lbl 1
:A+2(J-1)W→L
:A+2JW→R
:(L+R)/2→M
:L→X
:Y₁→L
:M→X
:Y₁→M
:R→X
:Y₁→R
:W(L+4M+R)/3+S→S
:IS>(J,D)
:Goto 1
:Disp "INTEGRAL ="
:Disp S
:End
```

6. MIDPOINT RULE

```
Prgm  :MIDPNT
:Disp"ENTER A,B,
AND N, NUMBER OF D
IVISIONS"
:Disp"ENTER A"
:Input A
:Disp "ENTER B"
:Input B
:Disp"ENTER N"
:Input N
:(B-A)/N→D
:A+D/2→X
:Y₁→S
:1→M
:Lbl 2
:X+D→X
:Y₁+S→S
:1+M→M
:If M<N
:Goto 2
:S*D→S
:Disp"APPROX IS"
:Disp S
:End
```

7.   LIMIT OF A SEQUENCE

```
Prgm:LIMSEQ                          :Goto 1
:12345→B                             :Lbl 7
:2^(-25)→E                           :Disp"LIMIT IS"
:1→X                                 :Disp A
:Lbl 1                               :Goto 9
:Y₁→A                                :Lbl 8
:If abs(A-B)<E                       :Disp"AFTER 2^3
:Goto 7                              0 TERMS NO LIMIT"
:A→B                                 :LBL 9
:2*X→X                               :End
:IfX>2^30
:Goto 8
```

8.   SUM OF A SERIES

```
Prgm  :SERIES                        :Disp"SUM IS"
:0→S                                 :Disp S
:.00000001→E                         :Goto 9
:1→X                                 :Lbl 8
:Lbl 1                               :Disp"AFTER 2^
:S+Y₁→K                              10 TERMS, NO LI
:If abs(S-K)<E                       MIT"
:Goto 7                              :Disp"Y₁="
:K→S                                 :Disp Y₁
:1+X→X                               :Disp "S="
:If X>2^(10)                         :Disp S
:Goto 8                              :Lbl 9
:Goto 1                              :End
:Lbl 7
```

A-2: PROGRAMS FOR THE TI-85

1. ZERO: THE BISECTION PROCESS

```
PROGRAM: ZERO
:Disp "CHOOSE A,B,SO"
:Disp "   F(A)F(B)<0"
:Lbl A
:Disp "ENTER A"
:Input A
:Disp "ENTER B"
:Input B
:.000000001→E
:A→x
:y1→P
:B→x
:If P*y1>0
:Goto H
:Lbl B
:(A+B)/2→M
:If abs(A-B)<E
:Goto I
:A→x
:y1→P
:M→x
:If P*y1>0
:Goto G
:M→B
:Goto B
:Lbl G
:M→A
:Goto B
:Lbl I
:Disp "ZERO AT"
:Disp M
:Disp "y1="
:Disp y1
:Goto E
:Lbl H
:Disp "F(A)F(B)>0,"
:Disp " CHOOSE AGAIN"
:Goto A
:Lbl E
:
```

2. LIMIT OF A FUNCTION

```
PROGRAM: LIMIT
:10000→P
:-10000→Q
:0→N
:.000000001→E
:Disp "ENTER C"
:Input C
:.1→D
:Lbl A
:C+D→x
:If abs(y1-P)<E
:Goto F
:y1→P
:C-D→x
:If abs(y1-Q)<E
:Goto G
:y1→Q
:D*.1→D
:1+N→N
:If N>12
:Goto E
:Goto A
:Lbl E
:Disp "NO LIMIT AFTER 12"
:Disp "APPROXIMATIONS"
:Goto I
:Lbl F
:Disp "RIGHT LIMIT"
:Disp P
:Lbl G
:Disp "LEFT LIMIT"
:Disp Q
:Lbl H
:If abs(P-Q)>100*E
:Goto D
:(P+Q)/2→W
:Disp "LIMIT IS"
:Disp W
:Goto I
:Lbl D
:Disp "RIGHT AND LEFT LIMITS"
:Disp " DIFFER OR ONE DOES"
:Disp " NOT EXIST, NO LIMIT"
:Lbl I
:
```

3.   NEWTON'S METHOD

```
 PROGRAM: NEWTON
:Disp "ENTER E"
:Input E
:Disp "ENTER X"
:Input x
:Lbl A
:x-(y1/y2)→A
:If abs(x-A)<E
:Goto B
:A→x
:Goto A
:Lbl B
:Disp "SOLUTION IS"
:Disp A
:
```

4.   TRAPEZOIDAL RULE

```
 PROGRAM: TRAP
:Disp "ENTER A,B AND N,"
:Disp "NUMBER OF DIVISIONS"
:Disp "ENTER A"
:INPUT A
:Disp "ENTER B"
:Input B
:Disp "ENTER N"
:Input N
:(B-A)/N→D
:A→x
:y1/2→K
:B→x
:y1/2+K→K
:A→x
:1→M
:Lbl B
:x+D→x
:y1+K→K
:1+M→M
:If M<N
:Goto B
:K*D→K
:Disp "APPROX IS"
:Disp K
:
```

5.   SIMPSON'S RULE

```
PROGRAM: SIMPSON
:FnOff
:Disp "LOWER LIMIT"
:Input A
:Disp "UPPER LIMIT"
:Input B
:Disp "N DIVISONS"
:Input D
:0→S
:(B-A)/2D→W
:1→J
:Lbl A
:A+2(J-1)W→L
:A+2*J*W→R
:(L+R)/2→M
:L→x
:y1→L
:M→x
:y1→M
:R→x
:y1→R
:W(L+4M+R)/3+S→S
:IS>(J,D)
:Goto A
:Disp "INTEGRAL ="
:Disp S
:
```

6.   MIDPOINT RULE

```
 PROGRAM: MIDPNT
:Disp "ENTER A, B, AND N,"
:Disp "NUMBER OF DIVISIONS"
:Disp "ENTER A"
:Input A
:Disp "ENTER B"
:Input B
:Disp "ENTER N"
:INPUT N
:(B-A)/N→D
:A+D/2→X
:y1→S
:1→M
:Lbl B
:x+D→x
::y1+S→S
:1+M→M
:If M<N
:Goto B
:S*/d→S
:Disp "APPROX IS"
:Disp S
:
```

7. LIMIT OF A SEQUENCE

```
 PROGRAM: LIMSEQ
:0→B
:2^(-25)→E
:1→x
:Lbl A
:y1→A
:If  abs(A-B)<E
:Goto G
:2*x→x
:If  x>2^30
:Goto H
:Goto A
:Lbl G
:Disp "LIMIT IS"
:Disp A
:Goto I
:Lbl H
:Disp "AFTER 2^30 TERMS"
:Disp " NOT LIMIT"
:Lbl I
:
```

8.   **SUM OF A SERIES**

```
 PROGRAM: SERIES
:0→S
:.00000001→E
:1→x
:Lbl A
:S+y1→K
If abs(S-K)<E
:Goto G
:K→S
:1+x→x
:If x>2^(10)
:Goto H
:Goto A
:Lbl G
:Disp "SUM IS"
:Disp S
:Goto I
:Lbl H
:Disp "AFTER 2^(10) TERMS,"
:Disp " NO LIMIT"
:Disp "y1="
:Disp y1
:Disp "S="
:Disp S
:Lbl I
:
```

A-3: PROGRAMS FOR THE SHARP

1. ZERO: THE BISECTION PROCESS

```
zero
_____REAL
Goto start
Label equation
y=(function)
Return
Label start
Print "choose a,b, so f(a)f(b)<0
Label 1
Input a
Input b
e=.000000001
x=a
Gosub equation
ya=y
x=b
Gosub equation
yb=y
If ya*yb>0 Goto 8
Label 2
m=(a+b)/2
x=a
Gosub equation
ya=y
x=m
Gosub equation
ym=y
If abs(a-b)<e Goto 9
If ya*ym>0 Goto 7
b=m
Goto 2
Label 7
a=m
Goto 2
Label 9
Print "Zero at
Print m
Print y
Goto 5
Label 8
Print "f(a)f(b)>0; choose again
Goto 1
Label 5
End
```

2.   LIMIT OF A FUNCTION

```
limit
―――――――――REAL
Goto start
Label equation
y=(function)
Return
Label start
p=10000
q=-10000
n=0
e=.000000001
d=.1
Label 1
x=c+d
Gosub equation
If abs(y-p)<e Goto 6
p=y
x=c-d
Gosub equation
If abs(y-q)<e Goto 7
q=y
d=d*.1
n=n+1
If n>12 Goto 5
Goto 1
Label 5
Print "after 12 approximations no limit
Goto 9
Label 6
Print "right  limit
Print p
Label 7
Print "left  limit
Print  8
If abs (p-q)>100*e Goto 4
w=(p+q)/2
Print "limit  is
Print  w
Goto 9
Label 4
Print "right and left limits differ or one does not exist, no lim
Label 9
End
```

3.   NEWTON'S METHOD

```
newton
_____REAL
Goto start
Label equation
y=  (function)
d=(derivative)
Return
Label start
Input e
Input x
Label 1
Gosub equation
a=x-(y/d)
If abs(x-a)<e Goto 2
x=a
Goto 1
Label 2
Print "solution is
Print a
End
```

4.   TRAPEZOIDAL RULE

```
trap
_____REAL
Goto start
Label equation
y=(function)
Return
Label start
Print "enter a,b, and n, number of divisions
Input a
Input b
Input n
d=(b-a)/n
x=a
Gosub equation
k=y/2
x=b
Gosub equation
k=y/2+k
x=a
m=1
Label 2
x=x+d
Gosub equation
k=k+y
m=m+1
If m<n Goto 2
k=k*d
Print "approx is
Print k
End
```

5.   SIMPSON'S RULE

```
simpson
_____REAL
Goto start
Label equation
y=(function)
Return
Label start
Print "lower limit
Input a
Print "upper limit
Input b
Print "number of divisions
Input d
s=0
w=(b-a)/2d
j=1
Label 1
l=a+2(j-1)w
r=a+2j*w
m=(l+r)/2
x=1
Gosub equation
l=y
x=m
Gosub equation
m=y
x=r
Gosub equation
r=y
s=w(l+4m+r)/3+s
j=j+1
If j<=d Goto 1
Print "integral =
Print s
End
```

6.   MIDPOINT RULE

```
midpoint
_____REAL
Goto start
Label equation
y=(function)
Return
Label start
Print "enter a, b, and n, number of divisions
Input a
Input b
Input n
d=(b-a)/n
x=a+d/2
Gosub equation
s=y
m=1
Label 2
x=x+d
Gosub equation
s=s+y
m=m+1
If m<n Goto 2
s=s*d
Print "approx is
Print s
End
```

7 LIMIT OF A SEQUENCE

```
limseq
_____REAL
Goto start
Label equation
y=(function)
Return
Label start
b=0
e=2^(-25)
x=1
Label 1
Gosub equation
a=y
If abs(a-b)<e Goto 7
b=a
x=x*2
If x>2^30 Goto 8
Goto 1
Label 7
Print "limit is
Print a
Goto 9
Label 8
Print "after 2^30 terms no limit
Label 9
End
```

8. SUM OF A SERIES

```
series
————REAL
Goto start
Label equation
y=(function)
Return
Label start
s=0
e=.00000001
x=0
Label 1
Gosub equation
k=s+y
If abs(s-k)<e Goto 7
s=k
x=x+1
If x>2^(10) Goto 8
Goto 1
Label 7
Print "sum is
Print s
Goto 9
Label 8
Print "after 2^(10) terms, no limit
Print y
Print s
Label 9
End
```

SOLUTIONS

1. INTRODUCTION

1. The first and the last.

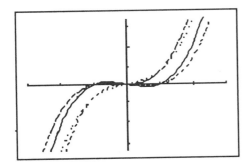

7. Yes; the second is the first shifted 1 unit to the right.

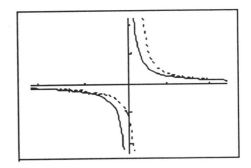

3. 0; ±2; 0,±1; ±2

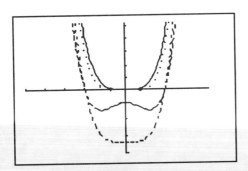

9. Yes; the second is the first shifted 1 unit right and 3 up.

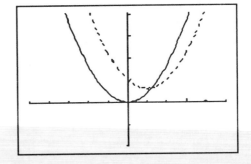

5. Only the last at x=-1. All get large as x approaches 0.

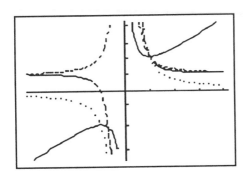

11. (a) Yes. The second can be written $y=(x+2)^2+1$.
     (b) Yes. The second can be written $y=(x-b/2)^2+c-b^2/4$.

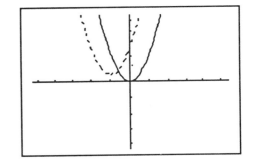

13.   (a)The y coordinates of the
       last three can be obtained
       by multiplying the y-coordinates
       of the first by a constant.
       (c) The y coordinates of af(x)
       can beobtained by multiplying
       the y-coordinates of f(x) by a.

    (b)

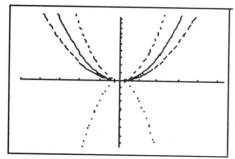

23.   neither

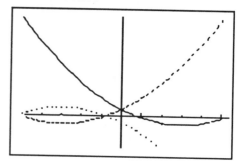

25.   odd

27.   odd

29.   neither

15.   .33

31.   even

17.   -3.338172844
       -.3926900267
       .7648536344
       12.96600924

33.   f is odd

19.   2.448048519

21.   .4204930342

2. ALGEBRAIC FUNCTIONS

1. 4; 2

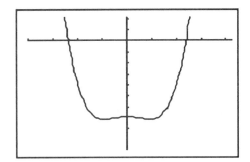

3. 6; 2

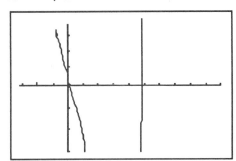

5. 2; 2

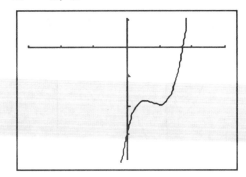

7. 2; 0

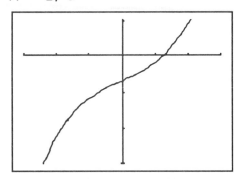

9.

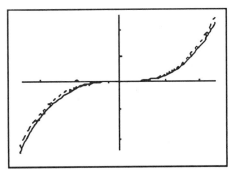

NOTATION:
   VA means vertical asymptote;
   HA means horizontal asymptote.

11. VA: x=2

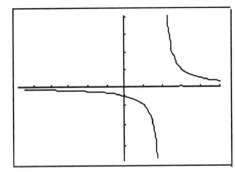

13. VA: x=-1

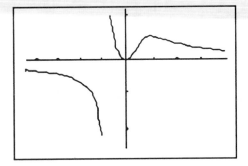

15.   VA: x=-1, x=1; HA: y=1

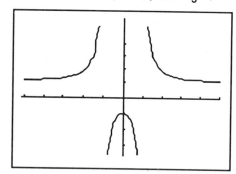

21.   y=x; y=-x

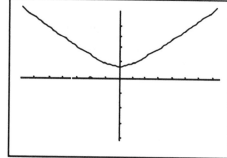

17.   VA: x=-1, x=1; HA: y=0

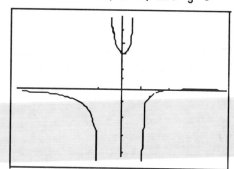

23.   x=2; x=-2

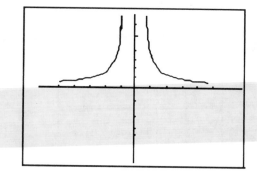

19.   VA: x=-1, x=1, x=2; HA: y=1

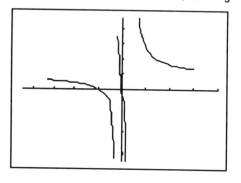

25.   VA: x=0 The function can be
       written as $y=x^2- 8/x$.
       For $-15 \le x \le -10$, the 8/x term
       is small.

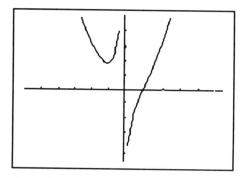

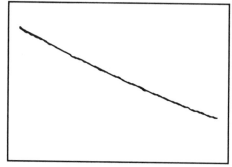

3. TRIGONOMETRIC FUNCTIONS

1.

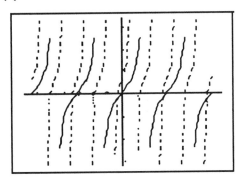

3.

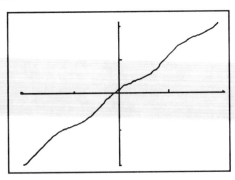

5.

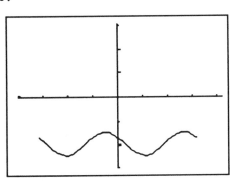

7.   yes

9.   no

11.   yes

13.   even function; need only
       compute the positive zeros.
       Their negatives will also
       be zeros.
       ±.3762879926
       ±1.181195621

15.   0, 1,
       -.6536531136
       -.1316129448
       1.056232684

17.   odd: sin, tan, csc, cot
       arcsin, arctan, arccsc
       even: cos, sec

19.   As x gets larger, $10/(x^2+1)$
       gets close to 0 and
       $\cos[10/(x^2+1)]$ gets close
       to 1.

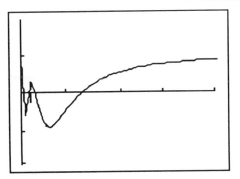

4.   LIMITS

(For 1 and 2, your answers for δ may be different from the ones given but may be just as correct; there is no one correct answer.)

1.   20; for ε=.1, let δ=.002
        for ε=.01, let δ=.0006

3.   .2222222222

5.   3

7.   2

9.   .9887710779

11.   4

13.   0

15.   .4

17.   -8.242640687,
        .2426406867

19. $f(x) = \dfrac{|x-4|}{x-4}$

$\lim\limits_{x \to 4^+} f(x) = 1$, $\lim\limits_{x \to 4^-} f(x) = -1$

but $\lim\limits_{x \to 4} |f(x)| = 1$

5. DERIVATIVES

1.

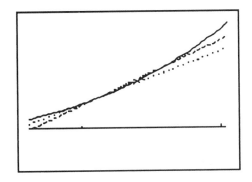

3.

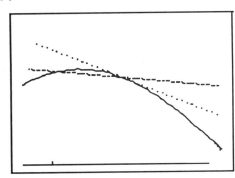

5.

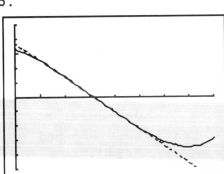

7.

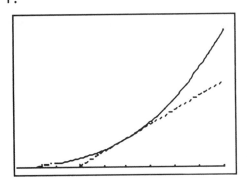

9. 1.547086268

11. 3.082207001

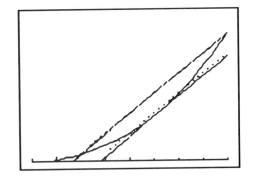

13. $6 \leq x \leq 10$, $10 \leq y \leq 20$

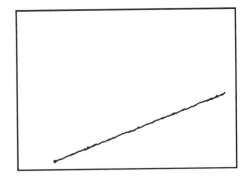

15. one such number is 2.277023269

17. (a) .7; (b) .7; (c) same; this is an application of L'Hopital's Rule

6. NUMERICAL DIFFERENTIATION

1.   -1.710124966
3.   1.959050199
5.   .724680279
7.   -.1636201945
9.   13.11304757

11.   1.553378373 with $\Delta x = .01$;
     1.553371199 from calculus
13. $y - .5525 = -.1417(x - 1.4)$
15.   Upward; .6784430159 ft/sec

7. MAXIMA AND MINIMA

1. $y = 11.432323$; $y = 4.8145790$
computed starting with RANGE
[-3.5.1.-20.20.5.1]; if you
start differently, your answers
may be slightly different

3.   the function has period 2π;
the max and min values on
$[0, 2\pi]$ are 3.3655804,
-1.835682, 1, -1.835682,
3.3655804, -4.561139,
5, -4.561139

5. $y = -201.6977541$ at
$x = -4.273752995$, loc min
$y = 2.487576111$ at
$x = -.1606735517$, loc max
$y = -42.95779074$ at
$x = 2.184426547$, loc min

7. $y = -1.76851387$ at
$x = -4.111414459$, loc max
$y = -3.434837933$ at
$x = -1.866208874$, loc min
$y = 9.52980847$ at
$x = 4.817516418$, loc max

9. $y = 1.549440034$ at
$x = .9477471332$: loc min

11. $y = 7.447890521$ at
$x = .3942752616$ and at
$x = 2.747317392$: loc max
$y = -3.637520151$ at
$x = 3.850237958$ and
$x = 5.574540004$: loc min
$y = -9$ at $x = \pi/2$: loc min
$y = 1$ at $x = 3\pi/2$: loc max

13.   y=-3.23831022 at
      x=-1.409819644, loc min

17.   x=74.35
      About 292.1 feet
      of water line

15. $2.50 yields $1,168.90
      $2.55 yields $1,173.01
      $2.60 yields $1173.49
      The price of $2.50 will
      probably be the best since
      the yields are about the
      same.

19.   Minimum distance is
      .5894389623

21.   x≈1.22625

8.   INCREASING AND DECREASING FUNCTIONS

1.   inc:(-∞,-.097]
      dec:[-.097,3.426]
      inc:[3.426,∞)

11.   inc:[0,1.977383029]
      dec:[1.977383029,3.837467107]
      inc:[3.837467107,2π]

3.   dec:(-∞,-4.64]
      inc:[-4.64,-2)
      inc:(-2,-.86]
      dec:[-.86,2)
      dec:(2,∞)

13.   inc:(-∞,-$\sqrt{19}$),inc:(-$\sqrt{19}$,0]
      dec:[0,$\sqrt{19}$),dec:($\sqrt{19}$,∞)
      You don't need the program
      here; just look at the graph
      and the denominator.

5.   dec for x≤0,
      inc for 0≤x

15.   Inc on [0,.795969903] and
      [2.316556045,∞)

7.   inc:(-∞,-.92744332278],dec:
      [-.92744332278,3.594109995]
      inc:[3.594109995,∞)

17.   No: Since, for example, we
      must have f(-2)=f(2), we
      cannot have f(-2)<f(2).

9.   inc:[0,.6368777459],dec:
      [.6368777459,1.457517354]
      inc:[1.457517354,2.731272848]
      dec:[2.731272848,3.551912459]
      inc:[3.551912459,4.82566795]
      dec:[4.82566795,5.646307562]
      inc:[5.646307562,2π]

9. CONCAVITY

1. dn:(-∞,-1.166666667]
    op:[-1.166666667,∞)

3. up:(-∞,-1.527525232]
    dn:[-1.527525232,1.527525232]
    up:[1.527525232,∞)

5. up:[0,.7599276495]
    dn:[.7599276495,1.997222358]
    up:[1.997222358,3.416072874]
    dn:[3.416072874,4.909501701]
    up:[4.909501701,2π]

7. up:[-1,-.5773502695]
    dn:[-.5773502695,.5773502695]
    up:[.5773502695,1].
    (even function)

9. up:[0,.3747344329]
    dn:[.3747344329,2.766859221]
    up:[2.766858221,2π]

11. dn:[0,.8057465555
    up:[.805746555,π/2]
    dn:[π/2,π]

13. up:(-∞,-1.358898944], dn:
    [-1.358898944,7.358898944]
    up:[7.35889844,∞)
    The second derivative
    completely determines
    concavity.

15. The second derivative of a
    polynomial of odd degree at
    least 3 is of odd degree and
    takes both positive and negative
    values. The concavity must
    change at least once.

10. NEWTON'S METHOD

1. 0
3. 1.05
5. .571
7. -3.40919, -1.30320,
    2.87400, 4.97998
9. 0, 3.03504

11. 1.51059204
    The calculator times are
    too close to tell the
    difference. However,
    ZERO does not require
    us to compute y'.

11. NUMERICAL INTEGRATION: INTRODUCTION

1. 27.00439453, 27

3. .8554752381

5. 0

7. .69331624389,
   2.080421687,
   2.777123955;
   it is 4 times
   the first

9. 18.36306887

11. 5.5355299296

13. -.19, -.32, -.40

15. .9315421293

17. see Appendix

19. .6628676

21. -27.733901367,
    -27.74725342,
    -27.74931335,
    -27.74982834.
    -27.75 with calculus

12. INVERSE FUNCTIONS

1. inverse exists

3. no inverse

5. 5.1694

7. $x=\sqrt{y/(1-y)}$; $x=-\sqrt{y/(1-y)}$

   -.65526316, .65526316

9. .0942003855

11. .184600494

13. $y=1\pm\sqrt{x-4}$
    -.2886751347,
    .2886751347

15. .1843024446, -.1843024446

13. LOGARITHMIC AND EXPONENTIAL FUNCTIONS

1. (a) 4.17438727;
     $\ln(xy) = \ln x + \ln y$

   (b) 3.295836866;
     $\ln(x/y) = \ln x - \ln y$

   (c) 9.406482648;
     $\ln(x^n) = n \ln x$

   (d) 1096.633158;
     $a^x a^y = a^{x+y}$

   (e) 1.9; $\ln(e^x) = x$

   (f) 5.7; $e^{\ln x} = x$

5. For $x>0$: $a^x < b^x$.
   For $x<0$: $a^x > b^x$.

7. Same: $b^{-x} = (b^{-1})^x = (1/b)^x$.

3. For $x<0$: $(.2)^x > (.3)^x > (.8)^x$
   For $x>0$: $(.2)^x < (.3)^x < (.8)^x$

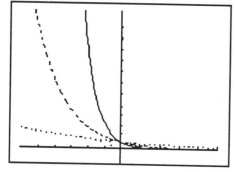

9. Same: $\ln(ab) = \ln a + \ln b$

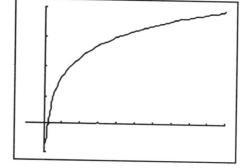

11. Same: ln(a/b)=lna - lnb

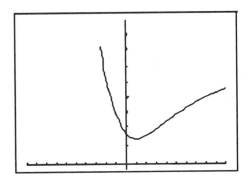

13. No: 2+sin x>0 and
     ln(2+sin x) exists.

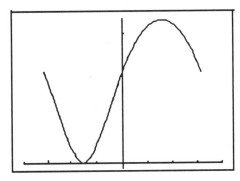

15. Yes, the function
     x^2 + cox x + 4 is even

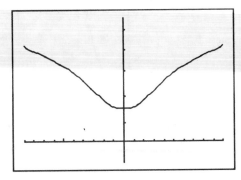

17. c=1000, m=.2772588722,
     y=9189.59 for x=8

19. 277.33; 277.39

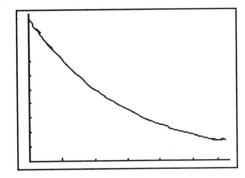

21. .0524966667

23. 3.3321

25. 1.469629767, 2.939259537
     First is half second because
     $\ln\sqrt{a} = \ln(a)^{1/2} = (1/2)\ln a$

27. loc min at x=.456961066,
     y=2.146720243

14   HYPERBOLIC FUNCTIONS

1.

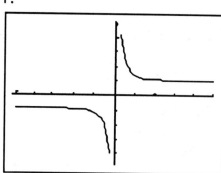

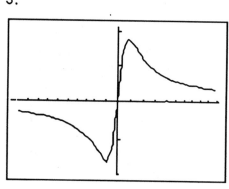

3.

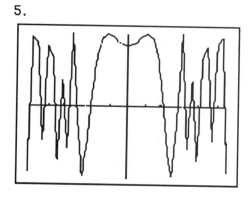

5.

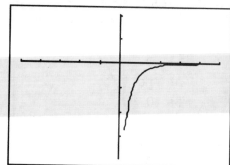

7.

9. e^x = cosh x + sinh x

e^{-x} = cosh x − sinh x

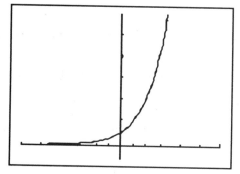

11.   Yes, they are the same

13.   .419065

15.   odd; odd

17.   y=1.87594988 at
x=.2462517112; loc min

19.   10.0806

15.   NUMERICAL INTEGRATION: APPLICATIONS

1.   .15883   11.   5.09602
3.   9.64746   13.   3.87469
5.   80.53840   15.   9.43048
7.   67.74263   17.   7290.2910
9.   71,792.0714   19.   -.99459496

16.   SEQUENCES AND SERIES

1.   (a)   1.4   7.   .58198
      (b)   0   9,   yes; .6666667
      (c)   no liit
      (d)   0
3.   (a)   2.71828 (actually e)
      (b)   1.27324
      (c)   0   11. $e^{\sin(\pi n)}=e^0=1$ for all
      (d)   no limit   integers n but $e^{\sin(\pi x)}\neq 1$
5.   .25   for x not an integer.

17.   TAYLOR POLYNOMIALS

1. $P_6(x)=1-(1/2)x^2+(1/24)x^4$ 3. $P_5(x)=1+x+(1/2)x^2+(1/6)x^3$

$-(1/720)x^6$ $+(1/24)x^4+(1/120)x^5$
                                                  polynomial:-1.221402667
                                                  calculator:-1.221402758

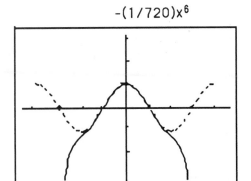   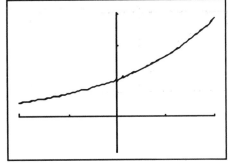

5. $P_4(x) = 1 + 2(x-\pi/4) + 2(x-\pi/4)^2$

   $+2(x-\pi/4)^3 + (7/3)(x-\pi/4)^4$
   1.546027015; within .01

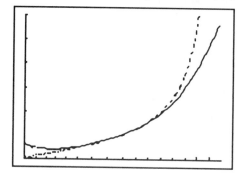

9. $P_5(x) = 1 - x + (1/2)x^2 - (1/6)x^3$

   $+(1/24)x^4 - (1/120)x^5$
   .0017

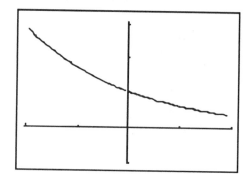

7. $P_6(x) = (1 - (x-1) + (x-1)^2 - (x-1)^3$

   $+(x-1)^4 - (x-1)^5$
   .0009145 ON [.5, 1.5]

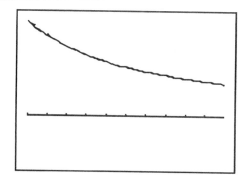

11.   The polynomial is the function.

13.   The derivative of the answer to Example 2 is the negative of the answer to Exercise 6. The derivative of cos x is -sin x.

15. $P_5(x) = (1/2)[e^x - e^{-x}]$

   $= x + (1/6)x^3 + (1/120)x^5$
   Sinh x is the odd part of e^x.

18. CONIC SECTIONS

1.

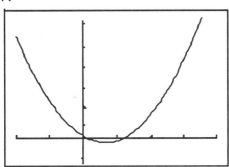

3.

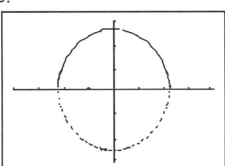

5.

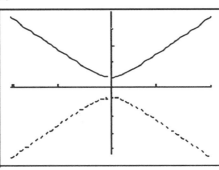

7.

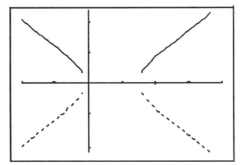

9.

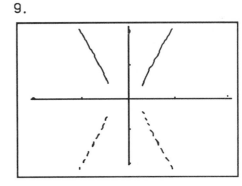

11. The second is the first shifted 3 units to the right and down 4.5.

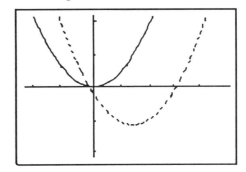

13. The second is the first with x and y coordinates exchanged.

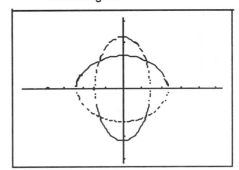

15. -1.000000139 with Δx=.001
    Actually it is -1.

17. 33.15242

19. 2.234837317
    2.868193819

19. PARAMETRIC EQUATIONS AND POLAR COORDINATES

1.

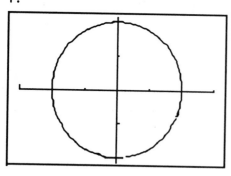

7. $x+y=\pi/2$ because
    $$\sin^{-1}t + \cos^{-1}t = \pi/2$$

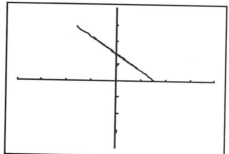

3. It is the graph of
    $y=1/x$ for $x>0$ since
    $$e^{-t}=1/e^{t}$$

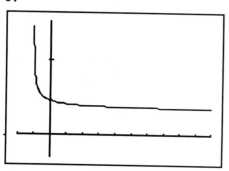

9. It is a line:
    $$y=(c/a)(x-b)+d$$

5.

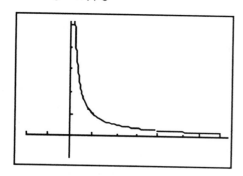

11.

13. An ellipse. It is a
     circle if a=b.

15. circle; radius a

17. (a)

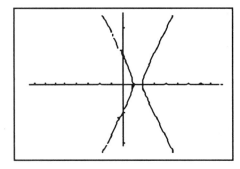

(b)

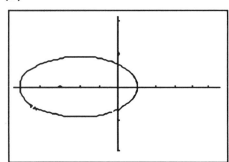

19. 98534.5

21.

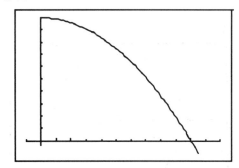

23.

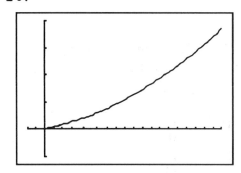

25.

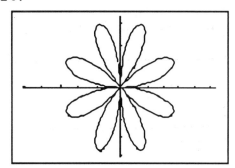

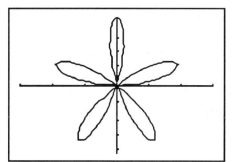

27.(a)

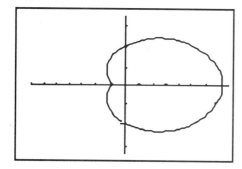

(b)

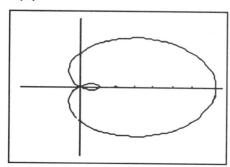

29.(a)

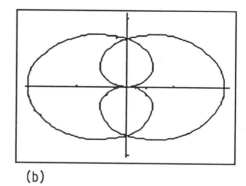

(b)

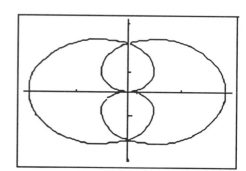

31. The second has the
     same shape as the first;
     $\cos \theta = \sin(\pi/2 - \theta)$

20. CURVE FITTING

1. $y=2.1634492.651969327)^x$

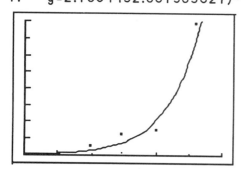

3. $y=3.202643172+1.572687225x$

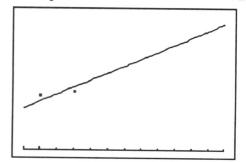

5. $y=5.471962617-.8691588785x$

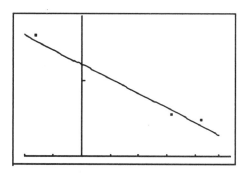

7.   Approximately
$y=184-60\ln x$

9. $y=14883(1.019)^x$

11.   East: $y=.618-.232x$
West: $y=.584-.047\ln x$
The east percentage
decreases faster.

13.   Results will vary with
data used.

15.   The regression numbers
are the same as before for
linear and exponential.
All that has changed are
the intercepts.

INDEX